A
DSP
Primer

with Applications to
Digital Audio
and
Computer Music

Ken Steiglitz

Department of Computer Science

Princeton University

Addison-Wesley Publishing Company, Inc.

Menlo Park, California • Reading, Massachusetts
New York • Don Mills, Ontario • Harlow, U.K. • Amsterdam
Bonn • Paris • Milan • Madrid • Sydney • Singapore • Tokyo
Seoul • Taipei • Mexico City • San Juan, Puerto Rico

Acquisitions Editor: Tim Cox
Executive Editor: Dan Joraanstad
Projects Manager: Ray Kanarr
Production Coordinator: Deneen Celecia
Cover Design: Yvo Riezebos Design
Text Design: Arthur Ogawa, TEX Consultants
Copy Editing: Elizabeth Gehrman
Proofreader: Joe Ruddick
Marketing Manager: Mary Tudor
Composition Services: Ken Steiglitz
Manufacturing Coordinator II: Janet Weaver

Instructional Material Disclaimer
The examples presented in this book have been included for their instructional value. They have been tested with care but are not guaranteed for any particular purpose. The publisher does not offer any warranties or representations, nor does it accept any liabilities with respect to the programs or examples.

Library of Congress Cataloging-in-Publication Data
```
Steiglitz, Ken, 1939-
   A digital signal processing primer, with applications to digi-
tal audio and computer music /
Ken Steiglitz. --1st ed.
     p.      cm.
1. Signal processing--digital techniques. 2. Computer sound
processing. 3. Computer music
   I. Title.
TK5102.9.S74     1995                        95-25182
621.382'2--dc20                                  CIP
```

ISBN 0-8053-1684-1

1 2 3 4 5 6 7 8 9 10–VG–99 98 97 96 95

Addison-Wesley Publishing Company
2725 Sand Hill Road
Menlo Park, CA 94025-7092

Contents

To my mom, Sadie Steiglitz, who will read it
during the commercials

Preface

Using computer technology to store, change, and manufacture sounds and pictures — digital signal processing — is one of the most significant achievements of the late twentieth century. This book is an informal, and I hope friendly, introduction to the field, emphasizing digital audio and applications to computer music. It will tell you how DSP works, how to use it, and what the intuition is behind its basic ideas.

By keeping the mathematics simple and selecting topics carefully, I hope to reach a broad audience, including:

- beginning students of signal processing in engineering and computer science courses;

- composers of computer music and others who work with digital sound;

- World Wide Web and internet practitioners, who will be needing DSP more and more for multimedia applications;

- general readers with a background in science who want an introduction to the key ideas of modern digital signal processing.

We'll start with sine waves. They are found everywhere in our world and for a good reason: they arise in the very simplest vibrating physical systems. We'll see, in Chapter 1, that a sine wave can be viewed as a phasor, a point moving in a circle. This representation is used throughout the book, and makes it much easier to understand the frequency response of digital filters, aliasing, and other important frequency-domain concepts.

In the second chapter we'll see how sine waves also arise very naturally in more complicated systems — vibrating strings and organ pipes, for example — governed by the fundamental wave equation. This leads to the cornerstone of signal processing: the idea that all signals can be expressed as sums of sine waves. From there we take up sampling and the simplest digital filters, then continue to Fourier series, the FFT algorithm, practical spectrum measurement, the z-transform, and the basics of the most useful digital filter design algorithms.

The final chapter is a tour of some important applications, including the CD player, FM synthesis, and the phase vocoder.

At several points I return to ideas to develop them more fully. For example, the important problem of aliasing is treated first in Chapter 3, then in greater depth in Chapter 11. Similarly, digital filtering is reexamined several times with increasing sophistication. This is why you should read this book from the beginning to the end. Not all books are meant to be read that way, but this one definitely is.

Some comments about mechanics: All references to figures and equations refer to the current chapter unless stated otherwise. Absolutely fundamental results are enclosed in boxes. Each chapter ends with a Notes section, which includes historical comments and references to more advanced books and papers, and a set of problems. Read the problems over, even if you don't work them the first time around. They aren't drill exercises, but instead mention generalizations, improvements, and wrinkles you will encounter in practice or in more advanced work. A few problems suggest computer experiments. If you have access to a practical signal-processing laboratory, use it. Hearing is believing.

Many people helped me with this book. First I thank my wife Sandy, who supports me in all that I do, and who helped me immeasurably by just being.

For his generous help, both tangible and intangible, I am indebted to Paul Lansky, professor of music and composer at Princeton. The course on computer music that we teach together was the original stimulus for this book.

I am indebted to many others in many ways. Perry Cook, Julius Smith, Tim Snyder, and Richard Squier read drafts with critical acumen, and their comments significantly improved the result. And I also thank, for assistance of various flavors, Steve Beck, Jack Gelfand, Jim Kaiser, Brian Kernighan, Jim McClellan, Gakushi Nakamura, Matt Norcross, Chris Pirazzi, John Puterbaugh, Jim Roberts, and Dan Wallach.

Ken Steiglitz
Princeton, N.J.

CHAPTER 1

Tuning Forks, Phasors

1 Where to begin

We've reached the point where sound can be captured and reproduced almost perfectly. Furthermore, digital technology makes it possible to *preserve* what we have *absolutely perfectly*. To paraphrase the composer Paul Lansky, we can grab a piece of sound, play it, and play with it, without having to worry about it crumbling in our hands. It is simply a matter of having the sound stored in the form of bits, which can be remembered for eternity.

Perfect preservation is a revolutionary achievement. Film, still the most accurate medium for storing images, disintegrates in just a few decades. Old sound-storage media — shellac, vinyl, magnetic wire, magnetic tape — degrade quickly and significantly with use. But bits are bits. A bit-faithful transfer of a compact disc loses nothing.

The maturing technology for digitizing sound makes the computer an increasingly flexible instrument for creating music and speech by both generating and transforming sound. This opens up exciting possibilities. In theory the computer can produce any sound it is possible to hear. But to use the instrument with command we must understand the relationship between what happens inside the computer and what we hear. The main goal of this book is to give you the basic mathematical tools for understanding this relationship.

You should be able to follow everything we do here with first-year calculus and a bit of physics. I'll assume you know about derivatives, integrals, and infinite series, but not much more. When we need something more advanced or off the beaten path, we'll take some time to develop and explain it. This is especially true of the complex variables we use. Most students, even if they've studied that material at one time, have not really used it much, and need to review it from scratch. As far as physics goes, we'll get an amazing amount of mileage out of Newton's second law: *force = mass × acceleration*.

Another goal of mine, besides providing a basis for understanding, is to amaze you. The mathematical ideas are wonderful! Think of it: Any sound that anyone will ever hear can be broken down into a sum of sine waves. There is more. Any sound that anyone will ever hear can be broken down into a file of bits — on/off positions of switches. That sound can be stored, generated, and manipulated on the computer is a miraculous technological incarnation of these mathematical principles.

Whenever possible we will approach a subject with simple physical motivation. We all have a lot of experience with things that make sound, and I want to lean on that experience as much as possible. So we'll begin with the simplest mechanism I can think of for making a sound.

2   Simplest vibrations

One of the easiest ways to produce a sound with a clear pitch, one we might call "musical" (leaving aside singing, which is actually very complicated), is to hit a metal rod like a tine of a tuning fork. The tine vibrates and sets the air in motion. Why does the rod vibrate? What waveform is produced?

The answer comes from simple physics. The tine is deformed when it is struck. A force appears to restore it to its original shape, but it has inertia, overshoots, and is deformed in the opposite direction. This continues, but each time the tine overshoots a bit less, and the oscillation eventually dies out. While it's oscillating, it's pushing the air, and the pressure waves in the air reach our ears. It is the balance between the two factors — the force that tends to restore the tine to equilibrium, and the inertia that tends to make it overshoot — that determines the frequency of oscillation. We will see a mathematical expression of this balance later in this section.

Suppose we think of hitting a tine of a tuning fork. As shown in Fig. 2.1,[†] we measure the deformation, or displacement, of the tine with the variable x. To keep things as simple as possible, assume the force that tends to restore the tine to its original position is proportional to the displacement, and of course in the direction opposite to x. That is, when the tine is pushed in the positive x direction the force acts to pull it back — in the negative x direction. Therefore, $F = -kx$, where F is the restoring force and k is the proportionality constant relating F to the displacement.

Next we take into account Newton's second law of motion: When the restoring force acts on the tine, it produces an acceleration proportional to that force. This law is usually expressed as $F = ma$, where m is the mass of the tine, and a is the acceleration. We decided above that $F = -kx$, so we now have two expressions for the force, which must be equal:

$$F = ma = -kx \qquad (2.1)$$

[†] References to figures and equations throughout this book are within the current chapter unless otherwise stated.

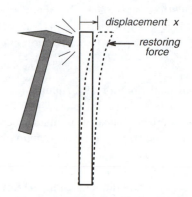

Fig. 2.1 Hitting a tine of a tuning fork. Small vibrations are sinusoidal.

Our goal is to learn exactly how the tine vibrates, but so far we seem to have derived only a relationship between its position and its acceleration. This turns out to be enough, however. Recall that the acceleration is just the second derivative of the displacement x with respect to the time variable t. Use this to rewrite Eq. 2.1 as

$$\frac{d^2 x}{dt^2} = -(k/m)x \tag{2.2}$$

In words, we are looking for a function $x(t)$ that is proportional to its second derivative. Furthermore, we know that the proportionality constant, $-(k/m)$, is a negative number. The solution is supplied from calculus, where we learn early on that

$$\frac{d}{dt}\sin(\omega t) = \omega\cos(\omega t) \tag{2.3}$$

and

$$\frac{d}{dt}\cos(\omega t) = -\omega\sin(\omega t) \tag{2.4}$$

where ω is the frequency of oscillation of the sine and cosine functions. Applying these differentiation operations one after the other gives

$$\frac{d^2}{dt^2}\sin(\omega t) = -\omega^2\sin(\omega t) \tag{2.5}$$

and

$$\frac{d^2}{dt^2}\cos(\omega t) = -\omega^2\cos(\omega t) \tag{2.6}$$

which shows that both $\sin(\omega t)$ and $\cos(\omega t)$ satisfy the equation for the tuning fork vibration, Eq. 2.2. In fact the functions $\sin(\omega t)$ and $\cos(\omega t)$ are really not much different from each other; the only difference is that one is a delayed version of the other. It doesn't matter when we consider time to begin, so the relative delay between the two is immaterial in this context. When it is not important to distinguish between sine and cosine we use the term *sinusoid*.

Equations 2.5 and 2.6 show that the sine and cosine functions have exactly the right shape to describe the vibration of the tuning fork. They also allow us to determine the frequency of vibration in terms of the physical constants k and m. Setting the constant $-(k/m)$ equal to the $-\omega^2$ term in Eq. 2.5 or 2.6, we get

$$\omega = \sqrt{k/m} \tag{2.7}$$

This sinusoidal vibration of the tuning fork is called *simple harmonic motion*. It arises in many contexts as the simplest kind of oscillatory behavior, or as the first approximation to more complicated oscillatory behavior. We see it in the motion of a pendulum, a stretched string — in any situation where there is a restoring force simply proportional to displacement.

Let's check to see if Eq. 2.7 makes intuitive sense. The variable ω, remember, is the frequency of oscillation; it tells us how fast the tuning fork vibrates. Every time the time t changes by $2\pi/\omega$, the argument ωt changes by 2π radians, and the sine or cosine goes through a complete cycle. Therefore the *period* of vibration is $2\pi/\omega$. The higher the frequency ω, the smaller the period. Recall that the variable m is the mass of the tuning-fork tine. All else being equal, Eq. 2.7 shows that the larger the mass, the lower the frequency. This confirms our experience. We all know that more massive objects vibrate at a slower rate than less massive objects. Larger tuning forks are the ones calibrated to the lower frequencies. Guitar strings corresponding to the lower pitches are thicker and heavier than those for the higher pitches.

Equation 2.7 also shows that the larger the restoring-force constant k, the higher the frequency of vibration. This also makes sense. A large value of k corresponds to a stiffer tine, which is analogous to a more tightly stretched string. We all know that tightening a guitar string raises the frequency of vibration. Not only does Eq. 2.7 tell us which way to expect the variation to go; it also tells us that the frequency of vibration is proportional not to the k/m ratio, but to the *square root* of that ratio. So, for example, to halve the frequency of a tuning fork tine we need to make it four times as massive. This proportionality of frequency of oscillation to the square root of a ratio between a measure of elasticity and a measure of inertia is a general phenomenon, and arises in many situations.

Return now to the observation that $\sin(\omega t)$ and $\cos(\omega t)$ are just shifted versions of each other. That is,

$$\cos(\omega t) = \sin(\omega t + \pi/2) \tag{2.8}$$

so sine and cosine differ by the fixed phase angle $\pi/2$, or a quarter-period. It is not hard to see that a sine or cosine with *any* phase angle satisfies Eq. 2.2. Just differentiate the function

$$x(t) = \sin(\omega t + \phi) \tag{2.9}$$

twice, and you get $-\omega^2 x(t)$ for any value of ϕ.

The relative phase angle ϕ is really arbitrary, and is determined only by the choice of time origin. To see this, replace t by $t + \tau$ in $\sin(\omega t)$, resulting in

$$\sin[\omega(t + \tau)] = \sin(\omega t + \omega\tau) = \sin(\omega t + \phi) \tag{2.10}$$

which shows that a fixed time shift of τ results in a phase shift of $\phi = \omega\tau$.

A more mathematical way to express this is to say that the set of all sinusoids at a fixed frequency is *closed* under the operation of time shift. In this sense the "shape" of a sinusoid is the same regardless of when we observe it. Sinusoids at the same frequency are also closed under addition, and we'll see that in the next section.

3 Adding sinusoids

The important closure property we establish next is that adding two sinusoids of the same frequency, but not necessarily with the same phases or amplitudes, produces another sinusoid with that frequency.

It's worth taking a minute to think about this claim. It is not as obvious a property as the invariance of sinusoids under shifting that we just proved. Figure 3.1 shows an example. It is perhaps obvious that the sum goes up and down with the same frequency, but why is the shape precisely that of a sinusoid? In physical terms it means this: If we strike two tuning forks tuned to exactly the same frequency, at different times and with different forces, the resulting sound, which is determined by the sum of the two vibrational displacements, is in theory indistinguishable from the sound of *one* tuning fork.

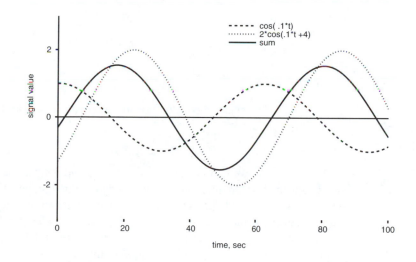

Fig. 3.1 Adding two sinusoids of the same frequency. The result is a third sinusoid of the same frequency.

The brute force way to show this would be to start with the sum

$$a_1 \cos(\omega t + \phi_1) + a_2 \cos(\omega t + \phi_2) \tag{3.1}$$

where a_1 and a_2 are arbitrary constants, and do some messy algebra. We would use the formulas for cosine and sine of sums, namely

$$\cos(\theta + \phi) = \cos\theta \, \cos\phi - \sin\theta \, \sin\phi,$$

$$\sin(\theta + \phi) = \sin\theta \, \cos\phi + \cos\theta \, \sin\phi \qquad (3.2)$$

plus a few other tricks, but when we were done we wouldn't really have gained much insight. Instead we will develop a better way of thinking about sinusoids, a way based on the circle.

After all, the first time you're likely to have seen a sinusoid is in trigonometry, and its definition is in terms of a right triangle. But a right triangle can be considered a way of projecting a point on a circle to a point on an axis, as shown in Fig. 3.2. We can therefore think of the cosine as the projection onto the x-axis of a point moving around the unit circle at a constant speed. Actually, the speed is simply ω radians per second, and it takes $2\pi/\omega$ seconds to complete one revolution, or *period*. In the same way, the sine wave can be thought of as the projection onto the y-axis. Figure 3.3 illustrates this; notice that I've been careful to start the cosine with the value 1 at $t = 0$ and the sine with the value 0 at that time.

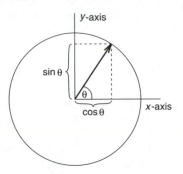

Fig. 3.2 Definition of cosine and sine as projections from the unit circle to the x- and y-axes.

From now on we can always think of a sinusoidal signal as a vector rotating at a steady speed in the plane, rather than a single-valued signal that goes up and down with a certain shape. If pressed, we can always point to the projection on the x-axis. But it's easier to think of a rotating clock-hand than some specially shaped curve. The position of the vector at the instant $t = 0$ tells us the relative phase of the sinusoid, what we called the angle ϕ, and the length of the vector tells us the size, or *magnitude*, of the sinusoid.

Now consider what happens when we add two sinusoids of the same frequency. This is the same as adding two force vectors in physics: we add the x parts and y parts, as shown in Fig. 3.4. The sum vector $u + v$ has an x-component that is the sum of the x-components of the addends u and v, and similarly for the y-component. The order of addition is immaterial, and the two possibilities form a parallelogram. That's why this law of vector addition is called the *parallelogram law*. Another way to put it is that the tail of the second vector is moved to the head of the first.

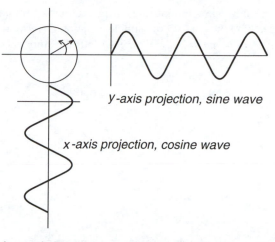
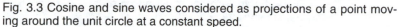

y-axis projection, sine wave

x-axis projection, cosine wave

Fig. 3.3 Cosine and sine waves considered as projections of a point moving around the unit circle at a constant speed.

Now we need to take into account the fact that the vectors representing sinusoids are actually rotating. But if the frequencies of the sinusoids are the same, the vectors rotate at the same speed, so the entire picture rotates as one piece. It is as if the vectors were made out of steel and the joints of the parallelogram were welded together. The result is that the parallelogram, together with the sum vector, also rotates at the same fixed speed, which shows that adding two sinusoids of the same frequency results in a third sinusoid of that frequency. Look again at Fig. 3.1. This shows projections onto the *x*-axis of the two components and the sum. Now maybe it doesn't seem so much of a miracle that the two special curves always add up to a third with exactly the same shape.

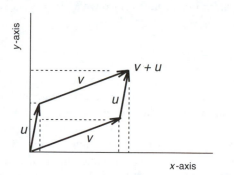

Fig. 3.4 Adding sinusoids by adding vectors.

4 Newton's second law

Our new view of sinusoids as projections of rotating vectors makes it even easier to see why they satisfy the simple equation governing the motion of the ideal tuning-fork tine, $F = ma = -kx$. Figure 4.1 tells the story geometrically.

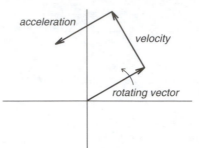

Fig. 4.1 Simple proof that sinusoids obey the equation of motion of the ideal tuning fork, Eq. 2.2.

The first derivative, or velocity, of a steadily rotating position vector is just another vector that is always at right angles to the rotating vector; in other words, it is tangent to the circle described by the rotating vector. This may sound mysterious at first, but it's really obvious. How does a rotating vector change when the time t increases by a small amount Δt? The vector's tip moves at right angles to the vector itself, so the new vector minus the old vector is tangent to the circle that the tip is tracing out. Put another way, the velocity vector points in the direction in which the tip of the rotating vector is moving, and is tangent to the circle it's tracing out.

The second derivative vector, or acceleration, has the same relationship to the velocity vector: it is always at right angles to it. Therefore, the acceleration vector is turned 180° with respect to the position vector. But this is just another way of saying that x, the position vector, and a, the acceleration vector, maintain themselves in opposite directions, which is also what $ma = -kx$ says.

5 Complex numbers

Complex numbers provide an elegant system for manipulating rotating vectors. The system will allow us to represent the geometric effects of common signal processing operations, like filtering, in algebraic form. The starting point is very simple. We represent the vector with x-axis component x and y-axis component y by the complex number $x + jy$. All complex numbers can always be broken down into this form; the part *without* the j factor is called the *real* part, and the part *with* the j factor the *imaginary* part. From now on we will call the x- and y-axes the *real-* and *imaginary-axes*, respectively.[†]

[†] Mathematicians use i, but electrical engineers use j to avoid confusion with the symbol for electrical current. Another vestigial trace of electrical engineering culture is the use of DC to mean zero frequency.

We can think of the j as meaning "rotate $+90°$" (counterclockwise). That is, we can think of multiplication by j as an *operation* of rotation in the plane. Two successive rotations by $+90°$ brings us to the negative real axis, so $j^2 = -1$. When viewed as a number, j of course plays the role of $\sqrt{-1}$. The geometrical viewpoint makes it clear that there's nothing mystical or imaginary about what might seem to be an impossible thing — a number whose square is -1.

Return now to our representation of sinusoids using rotating vectors. In terms of complex numbers, a complex sinusoid is simply

$$\cos(\omega t) + j\sin(\omega t) \tag{5.1}$$

This is just an algebraic way of representing Fig. 3.3.

If we were going to combine vectors only by adding them, the complex representation wouldn't give us any extra power. When we add complex numbers, we add the real parts and the imaginary parts, just as we add the x and y parts of force vectors. The extra zip we get from the complex-number representation comes when we *multiply* vectors. We will interpret any complex number as a rotation operator, just as we interpret j as the special operator that rotates by $+90°$. The next section is devoted to multiplication of complex numbers. That will put us in position to derive one of the most amazing formulas in all of mathematics.

6 Multiplying complex numbers

Let's pretend we've just invented complex numbers. How should we agree to multiply them? Obviously, we don't want to be arbitrary about it — we'd like multiplication to behave in the ways we're used to for real numbers. In particular, we want complex numbers with zero imaginary parts to behave just like real numbers. We also want multiplication to obey the usual commutative and associative rules.

Note that there is more than one way to define multiplication of two-component vectors. If you've studied mechanics you've seen the *cross product*. The cross product of vectors $\vec{x}$ and $\vec{y}$, denoted by $\vec{x} \times \vec{y}$, is another two-dimensional vector, but it is in fact at right angles to the plane of $\vec{x}$ and $\vec{y}$! To make matters more bizarre, the cross product is not commutative; that is, in general, $\vec{x} \times \vec{y} \neq \vec{y} \times \vec{x}$. We don't have any use for this multiplication now.

The way we'll define multiplication is to follow the rules of algebra blindly, using the usual distributive and commutative laws, and replacing j^2 by -1 whenever we want to. For example, to multiply the two complex numbers $x + jy$ and $v + jw$:

$$(x + jy) \cdot (v + jw) = (xv - yw) + j(xw + yv) \tag{6.1}$$

The $-yw$ term appears when j^2 is replaced by -1. This definition results in a complex multiplication operation that inherits the commutative and distributive properties of ordinary real multiplication.

Some computer scientists end their pictures of linked lists with the symbol for an electrical ground. One of my favorite vestigial symbols is the d for pence in English money — which is left over from the denarius of ancient Rome.

The real beauty of this definition is revealed when we think of complex numbers as vectors in *polar* form; that is, as vectors described by their lengths and angles. The length of the complex number $z = x + jy$, conventionally called its *magnitude* and written $|z|$, is just its Euclidean length in the plane, interpreting x and y as coordinates:

$$R = |z| = \sqrt{x^2 + y^2} \tag{6.2}$$

The angle it makes with the real axis, often called its *argument* or *arg*, or simply its *angle*, and written $\mathrm{ARG}(z)$, is just

$$\theta = \mathrm{ARG}(z) = \arctan(y/x) \tag{6.3}$$

We write the complex number z itself as $R\angle\theta$. (Read this to yourself as ''R at an angle θ.'') The complex number $(1+0j)$ is $1\angle0°$; the complex number $(0+1j)$ is $1\angle90°$.

To go back and forth between the $x + jy$ representation and the polar form $R\angle\theta$ is easy. The complex number $x + jy$ is a point on the circle of radius R, and from Fig. 3.2 we see that

$$x = R\cos\theta \tag{6.4}$$

and

$$y = R\sin\theta \tag{6.5}$$

In the other direction,

$$R = \sqrt{x^2 + y^2} \tag{6.6}$$

and

$$\theta = \arctan(y/x) \tag{6.7}$$

Now consider what multiplication should do in terms of the polar form. To be consistent with usual multiplication of real numbers, we want

$$R_1\angle0° \cdot R_2\angle0° = R_1 \cdot R_2\angle0° \tag{6.8}$$

This suggests that in general the magnitude of the product of any two complex numbers should be equal to the product of the two magnitudes.

Consider next the angle of a product of complex numbers. We've already said that we want to interpret the complex number j as rotation by 90°. This means we want multiplication by $j = 1\angle90°$ to add 90° to the angle of any complex number, but to leave the magnitude unchanged. This suggests that multiplication in general should result in adding the angles of the two complex numbers involved.

We have just given at least a plausibility argument for the following property of complex multiplication: *Multiply the magnitudes and add the angles.* That is, the product of $R_1\angle\theta_1$ and $R_2\angle\theta_2$ is $R_1 R_2\angle(\theta_1+\theta_2)$.

We now should verify this fact, given that we define multiplication as in Eq. 6.1. Verification is just a matter of checking the algebra. Let $x + jy = R_1\angle\theta_1$ and $v + jw = R_2\angle\theta_2$. Replacing x by $R_1\cos\theta_1$, y by $R_1\sin\theta_1$, and so forth in Eq. 6.1 gives the product $(x + jy) \cdot (v + jw)$ as

$$(R_1\cos\theta_1 R_2\cos\theta_2 - R_1\sin\theta_1 R_2\sin\theta_2) +$$
$$j(R_1\sin\theta_1 R_2\cos\theta_2 + R_1\cos\theta_1 R_2\sin\theta_2) \qquad (6.9)$$

The expressions in parentheses are familiar from Eq. 3.2. We circumvented them earlier to avoid some messy algebra, but now they come in handy, allowing us to rewrite this as

$$R_1 R_2\left[\cos(\theta_1 + \theta_2) + j\sin(\theta_1 + \theta_2)\right] \qquad (6.10)$$

which is just $R_1 R_2\angle(\theta_1 + \theta_2)$, exactly what we wanted to show. Multiplication does have the property that the magnitude of the product is the product of magnitudes, and the angle of the product is the sum of the angles.

We are now ready for Euler's formula, about which I can't say enough good things.

7 Euler's formula

The key fact we're looking for is that the rotating vector that represents a sinusoid is just a single fixed complex number raised to progressively higher and higher powers. That is, there's some fixed complex number, say W, that represents the rotating vector frozen at some angle; W^2 represents the vector at twice that angle, W^3 at three times that angle, and so forth. Not only that, but the vector W^ρ will represent a continuously rotating vector, where ρ is allowed to vary continuously over all possible real values, not just over integer values.

We concentrate our attention on a rotating vector of unit magnitude. More precisely we consider the function

$$E(\theta) = \cos\theta + j\sin\theta = 1\angle\theta \qquad (7.1)$$

which represents the vector at some arbitrary angle θ. From this we can find the derivative of $E(\theta)$ with respect to θ directly,

$$\frac{dE(\theta)}{d\theta} = -\sin\theta + j\cos\theta \qquad (7.2)$$

Notice next that the effect of the differentiation was simply to multiply $\cos\theta + j\sin\theta$ by j. In other words, we have derived the following simple property of $E(\theta)$:

$$\frac{dE(\theta)}{d\theta} = jE(\theta) \qquad (7.3)$$

We know that only the exponential function obeys this simple law. In general, the derivative of the function $e^{\alpha\theta}$ with respect to θ is α times the function, no matter what the value of α. This must be true even if α is complex, as it is in this case. In fact, $\alpha = j$ and the function we are looking for is

$$E(\theta) = e^{j\theta} \qquad (7.4)$$

This relation, written out in full,

$$\cos\theta + j\sin\theta = e^{j\theta} \qquad (7.5)$$

is called *Euler's formula*, after the Swiss mathematician Leonhard Euler (1707–1783). It is one of the most remarkable formulas in all of mathematics. It fulfills the promise above that the rotating vector arising from simple harmonic motion can be represented as a fixed complex number raised to higher and higher powers, and tells what raising a number to a complex power must mean. We will use it continually as we learn more about complicated signals and how to manipulate them.

Euler's formula ties together, in one compact embrace, the five best numbers in the universe, namely 0, 1, π, e, and j. To see this, just set $\theta = \pi$ in Eq. 7.5 and rearrange slightly (for aesthetic effect):

$$e^{j\pi} + 1 = 0 \qquad (7.6)$$

Not only that, but Eq. 7.6 uses, also exactly once each, the three basic operations of addition, multiplication, and exponentiation — and the equality relation. One of everything!

Euler's formula gives us a satisfying interpretation of the rule for multiplying complex numbers that we derived in the previous section. A complex number z with magnitude R and angle θ, $R\angle\theta$, can also be written $Re^{j\theta}$. The real part is $R\cos\theta$ and the imaginary part is $R\sin\theta$. Multiplying two complex numbers $z_1 = R_1\angle\theta_1$ and $z_2 = R_2\angle\theta_2$ can then be expressed as

$$z_1 \cdot z_2 = R_1 e^{j\theta_1} \cdot R_2 e^{j\theta_2} = R_1 R_2 e^{j(\theta_1 + \theta_2)} \qquad (7.7)$$

using the rule from the previous section: multiply their magnitudes and add their angles. This is an example of a general property that we expect from exponents, and that we'll use for complex numbers z, a, and b without further comment:

$$z^a \cdot z^b = z^{a+b} \qquad (7.8)$$

We'll also use the property

$$(z^a)^b = z^{ab} \qquad (7.9)$$

Here's a very important bit of notation that we'll use over and over again. Given any complex number $z = Re^{j\theta}$, its *complex conjugate* is defined to be

$$z^* = Re^{-j\theta} \qquad (7.10)$$

That is, z^* has the same magnitude as z, but appears in the complex plane at the negative of its angle. You can also look at it as an operation: to take the complex conjugate of a complex number, just replace j by $-j$ everywhere. Therefore, if $z = x + jy$ in terms of real and imaginary parts, its conjugate is $z^* = x - jy$. Geometrically, this means that z^* is the reflection of z in the real axis — it's as if that axis were a mirror. Points above the real axis are reflected below, and vice versa.

It's now easy to see that if we add a number and its complex conjugate, the imaginary parts cancel out, and the real parts add up. So if $z = x + jy$,

$$z + z^* = 2x = 2\,Real\,\{z\} \tag{7.11}$$

where we'll use the notation $Real$ to indicate the real part of a complex number. Similarly, if we subtract the conjugate of z from z, the real parts cancel, and

$$z - z^* = 2jy = 2j\,Imag\,\{z\} \tag{7.12}$$

where $Imag$ indicates the imaginary part.

What happens if we *multiply* a number times its conjugate? Euler's formula and Eq. 7.8 tell us that if $z = Re^{j\theta}$,

$$z \cdot z^* = Re^{j\theta} Re^{-j\theta} = R^2 = |z|^2 \tag{7.13}$$

This is a very convenient way to get the squared magnitude of a complex quantity.

By the way, the rotating vector derived earlier can now be written

$$e^{j\omega t} \tag{7.14}$$

We call such a signal a *phasor*, and regard it as a solution to the differential equation describing the vibrating tine of the tuning fork, Eq. 2.2. As discussed earlier, it is complex-valued, but if we want to use it to describe a real sound wave, we can always consider just the real part.

8 Tine as phasor

I can't resist taking a moment to point out that the piece of metal we've been hitting, the tuning-fork tine, can be made to vibrate as a phasor quite literally; that is, in a real circle. We can do this by striking it first in the x direction, and then in the y direction, perpendicular to the x direction.

To make this work we have to be careful to do it just right. First, we take care to strike the tine in both directions with equal intensity. Second, we strike it in the y direction precisely at the time when it has moved farthest in the positive x direction. Finally, we need to construct the tine so that its stiffness constant k is the same when moving in either the x or y direction.

Observing these precautions, suppose we first hit the tine in the positive x direction at the time corresponding to $\omega t = -\pi/2$. This means we get sinusoidal motion that has zero displacement at that time. The tine therefore vibrates in the x direction, and its displacement is described by

$$x(t) = \cos(\omega t) \tag{8.1}$$

We hit the tine a quarter-period early so that it will be farthest in the x direction at $t = 0$. We assume here for simplicity that we adjust the intensity of the strike so the amplitude of the oscillation is unity.

Next, strike the tine in the positive y direction at the time $t = 0$; that is, as planned, a quarter-period later, when the tine is fully deflected in the positive x direction. The vibration in the y direction is independent of that in the x direction, and is described by

$$y(t) = \sin(\omega t) \tag{8.2}$$

If we look at the tine from above it moves precisely in a circle, one revolution every $2\pi/\omega$ seconds. This is illustrated in Fig. 8.1. We have created a real phasor.

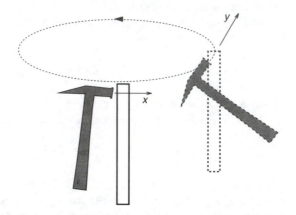

Fig. 8.1 Hitting a tine of a tuning fork twice, to start vibration first in the x direction and then in the y direction. As a result the tip of the tine moves in a circle.

Superpositions of oscillations in more than one direction, of which this is a simple example, can result in intricate patterns, especially if the oscillations have different frequencies. These patterns are called *Lissajous figures*, after the French physicist Jules Antoine Lissajous (1822–80). They make impressive pictures on oscilloscopes, and you can see them in older science-fiction films, for example at the beginning of *THX 1138*, a 1970 film directed and co-written by George Lucas.

9 Beats

We will conclude this chapter with an analysis of beats, a phenomenon familiar to anyone who experiments with sounds. Not only are beats interesting in themselves, but the analysis demonstrates quickly how useful our phasor representation is.

Suppose we strike two tuning forks that have frequencies of vibration that are close, but not identical. We know intuitively that the sinusoids from the two tuning forks shift in and out of phase with each other, first reenforcing, then destructively interfering with each other, as illustrated in Fig. 9.1. How do we represent this mathematically?

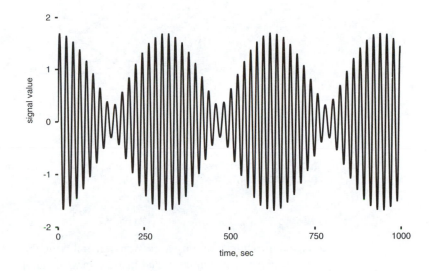

Fig. 9.1 Two sinusoids beating against each other. The exact function shown is sin(ωt) + 0.7*sin((ω + δ)t), where ω = 0.3157 radians per sec, and δ = 0.02 radians per sec.

We could write the two sinusoids in the following way

$$a_1\cos(\omega t) \; + \; a_2\cos((\omega + \delta)t) \tag{9.1}$$

where the frequencies differ by δ, which we assume is positive for the purposes of illustration. This does not show the beating phenomenon, and it takes a fair amount of messy algebra to put it in a form that does. But if we use phasors we can see what happens easily. Write the sum of two phasors as

$$a_1 e^{j\omega t} \; + \; a_2 e^{j(\omega + \delta)t} \tag{9.2}$$

Notice that Eq. 9.1 is just the real part of this. Now think of these phasors rotating in the complex plane. The second rotates at a rate δ faster than the first. The first vector begins in phase with the second, perfectly aligned, so the sum vector starts with length $|a_1 + a_2|$. As time progresses, the first vector drifts farther and farther behind the second, until it is 180° behind it and cancels it out, so that the sum vector shrinks in length to $|a_1 - a_2|$. The two phasors then gradually move back into phase, and so forth. The time that it takes for the two phasors to go through one such complete cycle is determined by the frequency δ. For example, if δ = 2π radians per sec (which corresponds to the frequency of 1 Hz), it takes one second to go from one relative null to the next. Figure 9.2 illustrates the relative motion of the two phasors in the complex plane.

An even more illuminating picture can be drawn in terms of phasors if we examine the expression for the magnitude of the sum phasor in Eq. 9.2. To do this, first factor out the common $e^{j\omega t}$ in that equation, yielding

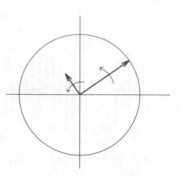

Fig. 9.2 Two phasors with different frequencies. They alternately line up and cancel out.

$$e^{j\omega t}\left[a_1 + a_2 e^{j\delta t}\right] \tag{9.3}$$

Next, take the magnitude of this expression, remembering that the magnitude of a product is the product of magnitudes, and that the magnitude of $e^{j\omega t}$ is always one. The result is

$$\left|a_1 + a_2 e^{j\delta t}\right| \tag{9.4}$$

This quantity is the magnitude of the vector that results from adding the constant real vector a_1 to the phasor with magnitude a_2 and frequency δ, as shown in Fig. 9.3. Remember that we removed the effect of the factor $e^{j\omega t}$ in Eq. 9.3 when we took the magnitude. That step canceled rotation of the entire configuration in Fig. 9.3 at a rate of $+\omega$ radians per sec, which, of course, doesn't affect the magnitude of the resultant sum vector. In effect, Fig. 9.3 shows motion relative to the rotating frame of reference determined by the original phasor at frequency ω.

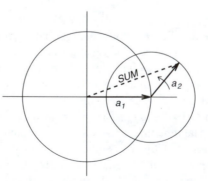

Fig 9.3 The complex vector representing the envelope of a beat signal, shown with a dashed line and labeled "SUM."

Figure 9.3 shows that the magnitude of the sum phasor is precisely the length of the link that connects the origin to the rim of the rotating wheel of radius a_2 centered at a_1. This link is much like the cam that drives a locomotive wheel. If we think of the sum of the two phasors as a complex vector that rotates with varying length and speed, we can define its varying length to be the *envelope* of the sum signal, and its varying angular speed to be its *frequency*. To emphasize the fact that this envelope and frequency are varying with time, we sometimes use the terms *instantaneous envelope* and *instantaneous frequency* (see Problems 9–13).

Notes

If you're an engineering or computer science student, I hope this book will whet your appetite for more, and that you'll go on to an upper-level course in digital signal processing, such as typically taught from the classic ''Oppenheim & Schafer'':

> A. V. Oppenheim and R. W. Schafer, *Digital Signal Processing*, Prentice-Hall, Englewood Cliffs, N.J., 1975.

A generation of students have learned digital signal processing from this source.

If you're a composer of computer music, you will want to add F. R. Moore's comprehensive book to your bookshelf:

> F. R. Moore, *Elements of Computer Music*, Prentice-Hall, Englewood Cliffs, N.J., 1990.

Moore describes in detail many of the tools used by composers, and provides practical and musical insights. The present volume should make Moore's book more accessible to you.

My experience is that, in general, colleges and universities teach calculus, but not algebra with complex numbers. Many of my students, even the best ones, tell me they haven't seen complex numbers since grammar school. That's why I start by going over complex arithmetic in some detail, but use first-year calculus freely throughout the book. My guess is that the calculus will be easier and more familiar to you, especially if you are a technical student. In most cases I try to provide the intuition behind what's going on, and I don't dwell on mathematical niceties. In fact some of the derivations are deceptively simple and shamefully unrigorous. What I'm after, and what is most useful to beginners, is intuition.

Turning to the material in the first chapter and confirming what I just said: Tuning forks aren't as simple as may be suggested by the analysis in Section 2; I just wanted to get started with simple harmonic motion as quickly as possible. First of all, there are two tines joined at the middle, so the fork is more like a full bar held at the center. The two tines interact. You can show this easily by touching one with your hand while the tuning fork is producing a tone. Both tines will stop vibrating. Second, the tines can vibrate in more complicated ways than suggested by the analysis. The picture we have for the simple harmonic motion is that the entire tine is swaying back and forth. But it can also be that the tip of the tine is moving one way while the middle part is moving the other. This mode of vibration results in what is called the *clang* tone, and can be excited by hitting a fork lightly with a metal object. More about modes of vibration in the next chapter.

If you want to learn more about the production and perception of musical sounds, the following book by the great master Hermann Helmholtz (1821–1894) is required reading:

> H. L. F. Helmholtz, *On the Sensations of Tone as a Physiological Basis for the Theory of Music*, Second English edition, A. J. Ellis, translator, Dover, New York, N.Y., 1954. (The first edition was published in German, 1863.)

It isn't the last word, but in many cases it's the first. Helmholtz is particularly well known to musicians for his contribution to the understanding of combination tones — the notes perceived when two sinusoids are sounded together. His key insight is the observation that slight nonlinearities in a musical instrument or in our ears explain why we hear sum and difference tones (see Problem 18).

Helmholtz had a special genius for making profound scientific observations with little or no apparatus. He worked wonders with a little dab of wax here, or a feather there. In connection with the beating of two sinusoids close in frequency, he observed in *Sensations*, "A little fluctuation in the pitch of the beating tone may then be remarked." This is the basis for Problem 13, which is solved in Helmholtz's Appendix XIV.

Problems

1. If we were mathematicians, we might want to state the closure-under-shift property of sinusoids with some precision. We might say that the class of functions

$$S_\omega = \{A\sin(\omega t + \phi)\}$$

for a fixed frequency ω and all real constants A and ϕ, is *closed* under the operation of time shift. This is a fancy way of saying that if we time-shift any member of the class, we get another member of the class. Find another class of time functions besides S_ω that is closed under the time-shift operation. Be precise in your definition of the class.

2. Is it true that the product of two members of the class S_ω is also a member of S_ω? That is, is the class S_ω closed under multiplication? Is it closed under division?

3. We demonstrated in Section 3 that the class S_ω is closed under addition. That is, we showed that adding two members of the class produces a third member of the class. Prove that adding any finite number of members of the class produces another member of the class. (That is, that the class is closed under *finite addition*.)

4. Is the class S_ω closed under addition of a countably infinite number of members? Think about this question. It will be answered in Chapter 3.

5. Suppose you simultaneously hit two tuning forks marked as being tuned to the same pitch. Name a couple of reasons you might *in practice* be able to distinguish the resulting sound from that of a single tuning fork, even though theory predicts that you wouldn't be able to.

6. Find two tuning forks that are marked as being tuned to the same pitch, digitize the sound of each being struck separately, and add the two notes on a computer. Do you get a single sinusoid?

7. Strike a tuning fork and hold it upright beside your ear. Then rotate it about its vertical axis. Explain why the loudness varies. Observe the angles at which the sound is softest and loudest.

8. Write a program that displays the Lissajous figures corresponding to vibrations in the x and y directions with different frequencies. Look especially at the patterns when the two frequencies are exactly, and then nearly, in the ratio of small integers. This used to be fun to do with an oscilloscope and a couple of signal generators, and if you have that equipment it's still an easy way to get the pictures and see them move as the two components drift in phase with respect to each other.

9. A student in a computer-music course decided he would generate a "chirp" signal — one that swept in frequency from ω_1 to ω_2 during the time interval $t = 0$ to $t = T$. To do this he generated the signal $\sin(\omega t)$, where

$$\omega = \omega_1 + \frac{t}{T}(\omega_2 - \omega_1)$$

The frequency variable ω does start at ω_1 and increases linearly to ω_2 as t increases from 0 to T.

Repeat his experiment and listen to the result. Does it have the expected pitch range? If you have a tuning fork, you can check the synthesized pitch by sweeping the frequency of the synthesized tone from a frequency below that of the tuning fork to one above it, and comparing the synthesized sound with that of the tuning fork. You can also measure its spectrum with a computer and a spectrum-analyzing program (about which more later).

Try to explain any discrepancy between the observed and the intended pitches. What will the signal sound like if we continue it at the fixed frequency ω_2 for $t > T$?

10. Derive an algebraic expression (in terms of real variables) for the envelope of the sum of two sinusoids, Eq. 9.1, in two ways. First, using the geometry in Fig. 9.3 (use the law of cosines); and second, using algebra and Eq. 9.4.

11. Prove that the actual waveform, Eq. 9.1, touches this envelope.

12. Find an equation of the form $g(\tau) = 0$ whose solutions are the points τ where the waveform touches the envelope. Try to solve it.

13. As mentioned at the end of Section 9, the *instantaneous frequency* of a complex signal is defined to be the rate of change of its angle. That is, if the angle of a signal is $\Theta(t)$, its instantaneous frequency is $d\Theta(t)/dt$. Consider the case of the phasor $e^{j\omega t}$, for example. Its magnitude is 1 and therefore its envelope is 1. Its angle is ωt, so its instantaneous frequency is ω, as we want.

Work out the instantaneous frequency of the sum of two sinusoids, Eq. 9.1. That is, derive as simple an algebraic expression as you can in terms of the real parameters a_1, a_2, ω, and δ. What are the smallest and largest values that the instantaneous frequency achieves? Are there values for the parameters for which you think you will be able to hear the variation in frequency? Synthesize the beat signal for these parameters and listen to it. Do you hear the predicted change in frequency?

14. Write a program that converts complex numbers from the form $x + jy$ to the polar form $R\angle\theta$. Write another that does the conversion in the opposite direction. Use *degree* as the unit of angle.

15. Derive Euler's formula using power series.

16. Get a tuning fork and measure the frequency of the tone it's meant to produce, let's say its nominal frequency. Then measure the frequency of the clang tone mentioned in the Notes. Is the clang-tone frequency an integral multiple of the nominal frequency? Does the clang tone die out faster or more slowly than the nominal frequency?

17. The beat signal shown in Fig. 9.1 is the result of adding two sinusoids that differ in frequency by $\delta = 0.02$ radians per second. What period does this correspond to in seconds? Check by measuring the figure.

18. A signal is produced by adding two sinusoids of frequencies ω_1 and ω_2, and squaring the sum. What frequency components are present in the result, and in what proportion?

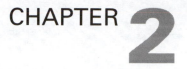

CHAPTER 2

Strings, Pipes, the Wave Equation

1 A distributed vibrating system

In the first chapter we considered the simplest kind of vibrating system, exemplified by a struck tine of a tuning fork, and showed how to describe its vibration mathematically. This led to phasors, a representation for sinusoids in the complex plane. I wanted to show you that sinusoids come up in the real world very naturally. In fact, we'll find out in this chapter that sinusoids are really fundamental building blocks out of which all sounds are composed. To see this we'll study the next simplest kinds of vibrating systems, beginning with the vibrating string.

The main difference between simple harmonic motion and the motion of a stretched string is that the string is distributed in space. That is, we no longer consider the motion of only one point, the tip of the tuning-fork tine, but we consider the motion of *infinitely many* points along the string. We will be looking for a description of the motion of the string as a function of *two* variables: the time, as before, but also position along the string. Let's denote that function by $y(x, t)$, where x is longitudinal position along the string, and y is the transverse displacement of the string with respect to its resting position. Figure 1.1 shows such a string; the x-axis represents the equilibrium position of the string, the flat line $y = 0$.

We're headed for an equation analogous to the differential equation in Eq. 2.2 of Chapter 1:

$$\frac{d^2x}{dt^2} = -(k/m)x \tag{1.1}$$

except now we have two variables to contend with. The displacement y of the string depends both on the position x along string, and the time t, and that's why we'll write it as $y(x, t)$. The derivatives in this more complicated situation are called *partial derivatives*, and the equation we will derive is called a *partial differential equation*.

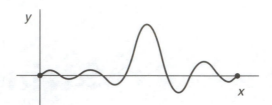

Fig. 1.1 A string stretched between two points, vibrating. The displacement y is a function of position x along the string and time t. Shown is a snapshot at a particular time.

There's really nothing very mysterious about this. It's just that we need to distinguish between the changes in the displacement y caused by variations in x and those due to changes in t. If we vary x but force t to remain constant, the resulting partial derivative is denoted by $\partial y/\partial x$. This represents the rate of change of y with respect to x, just as in the case of ordinary derivatives, except that we are being explicit about holding t fixed at some given value. Similarly, if we vary t but hold x constant, the result is $\partial y/\partial t$.

As a simple example consider the function

$$y(x,\ t)\ =\ e^{\alpha t}\sin(\omega x) \tag{1.2}$$

Then

$$\frac{\partial y}{\partial x}\ =\ \omega e^{\alpha t}\cos(\omega x) \tag{1.3}$$

and

$$\frac{\partial y}{\partial t}\ =\ \alpha e^{\alpha t}\sin(\omega x) \tag{1.4}$$

The particular partial differential equation we are about to derive is called the *wave equation*, and is one of the most fundamental in all of physics. It describes not only the motion of a vibrating string, but also a vast number of other situations, including the vibration of the air that enables sound to reach our ears.

2 The wave equation

The basic method of deriving the wave equation is straightforward. We consider a typical segment of the string, calculate the force on it, and apply Newton's second law. We then take the limit as the length of the segment goes to zero, and that's where we have to be careful in dealing with the partial derivatives.

Figure 2.1 shows a small piece of a vibrating stretched string, of length Δx. We assume that the tension on the string is P, and that the deformation of the string is small enough that we can ignore the change in tension in this segment caused by its deformation. We also assume that the string has uniform density ρ units of mass per unit length, so that the mass of the segment is $\rho\Delta x$.

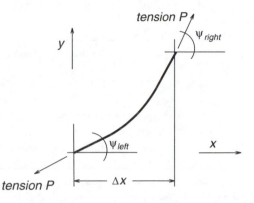

Fig. 2.1 An infinitesimal segment of a vibrating stretched string.

As shown, the string segment makes an angle ψ with the x-axis at the particular time and position x considered. The component of the tension in the vertical direction is $P\sin\psi$. Here's the tricky part. The angle ψ is not exactly the same at the two ends of the segment, because the string is curved. Therefore there is a difference in the vertical component at the two ends, given by

$$P\sin\psi_{left} - P\sin\psi_{right} \tag{2.1}$$

where ψ_{left} and ψ_{right} are the angles ψ at the left and right ends respectively. Write this as

$$P\Delta_x(\sin\psi) \tag{2.2}$$

where we use the notation $\Delta_x(\sin\psi)$ to denote the change in $\sin\psi$ as a result of changing x.

The next step, as mentioned, is to apply Newton's second law: the difference in the y components of the forces at the two ends must equal the mass of the segment times the acceleration in the y direction. The mass is $\rho\Delta x$, and the acceleration is $\partial^2 y/\partial t^2$. This gives

$$\rho\Delta x \frac{\partial^2 y}{\partial t^2} = P\Delta_x(\sin\psi) \tag{2.3}$$

Rearrange this by dividing by $\rho\Delta x$, yielding.

$$\frac{\partial^2 y}{\partial t^2} = (P/\rho)\frac{\Delta_x(\sin\psi)}{\Delta x} \tag{2.4}$$

We're now going to make the important assumption that the string displacement y, and hence the angle ψ, are very small. This is certainly true in the real world — a guitar string doesn't deviate much from its rest position to make sound. (The vertical scales in our figures are very exaggerated.) Mathematically, the assumption of small ψ means that $\sin\psi \approx \tan\psi \approx \partial y/\partial x$.

We now take the limit as Δx goes to zero. The expression $\Delta_x(\cdot)/\Delta x$ approaches $\partial(\cdot)/\partial x$, by definition — it's just the change of whatever is inside the parentheses divided by Δx, as $\Delta x \to 0$, keeping the time t constant. What's inside the parentheses approaches $\partial y/\partial x$, by the assumption in the preceding paragraph that ψ is small. The result is that the equation of motion of the string becomes

$$\frac{\partial^2 y}{\partial t^2} = (P/\rho)\,\frac{\partial^2 y}{\partial x^2} \tag{2.5}$$

Notice that the proportionality constant (P/ρ) is analogous to the constant (k/m) in the equation for simple harmonic motion, Eq. 1.1. The tension P is analogous to the stiffness constant k, being a measure of how resistant the string is to being displaced. The mass density ρ is directly analogous to the mass of the tuning-fork tine.

We can now get an important hint about the meaning of the proportionality constant (P/ρ). It must have the dimensions *distance-squared* over *time-squared*, as you can see easily from Eq. 2.5: formally replace y and x by *distance*, and t by *time*. So if we rewrite Eq. 2.5 as

$$\boxed{\frac{\partial^2 y}{\partial t^2} = c^2\,\frac{\partial^2 y}{\partial x^2}} \tag{2.6}$$

the constant c has the dimensions *distance* over *time*, or velocity. As we see in the next section, c really *is* a velocity. This is the *wave equation*, which, as we mentioned before, explains an enormous variety of wavelike phenomena in the physical universe.

The wave equation has an immediate intuitive interpretation. The right-hand side is proportional to the *curvature* of the string at any point x, and the left-hand side is proportional to how fast the string at that point is accelerating. If the curvature is positive, the string is $\cup$-shaped, and is accelerating upward (positive acceleration). If the curvature is negative, it is $\cap$-shaped and accelerating downward (negative acceleration). The sharper the bend of the string, the faster it is accelerating. If the string has zero curvature — that is, if it's in the shape of a straight line — its acceleration is zero, and it's therefore moving with constant velocity.

3 Motion of a vibrating string

We know from experience that if we suddenly shake the end of a string a wave will be generated that travels down the string. That's how a whip works. We see next that this is predicted by the wave equation. In fact, the result falls out of the wave equation immediately, with almost no effort.

Suppose then that we have shaken the end of the string, and produced a "bump" traveling to the right, as shown in Fig. 3.1. If the bump were in fact moving to the right, the deflection $y(x,\ t)$ of the string would be expressed mathematically by

$$y(x,\ t) = f(t - x/c) \tag{3.1}$$

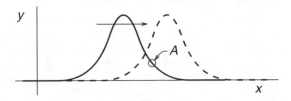

Fig. 3.1 A bump moving to the right on a string. The point at position A bobs up and then down as the bump passes.

where $f(\cdot)$ is a completely arbitrary function of one variable that represents the shape of the bump. To see this, just notice that if we increase t by Δt and x by Δx, the right-hand side of Eq. 3.1 remains unchanged, provided that $c\Delta t = \Delta x$. This means that the left-hand side, the deflection y, is the same at the later time $t + \Delta t$, provided we move to position $x + \Delta x$, where $\Delta x/\Delta t = c$. This is just another way of saying the shape of string moves to the right with speed c.

It is now easy to see that Eq. 3.1 always satisfies the wave equation, no matter what shape $f(\cdot)$ is. If we differentiate twice with respect to x, we get $(1/c^2)f''(t - x/c)$. If we differentiate twice with respect to t we get $f''(t - x/c)$, c^2 times the first result. This is exactly what the wave equation, Eq. 2.5, says.

If the wave is moving in the negative x direction, the deflection is of the form $y(x, t) = g(t + x/c)$, where $g(\cdot)$ is again any function of a single argument. The same procedure shows that this also satisfies the wave equation. In fact, any solution of the form

$$y(x, t) = f(t - x/c) + g(t + x/c) \tag{3.2}$$

will work, where the wave shapes $f(\cdot)$ and $g(\cdot)$ are completely arbitrary. This represents one wave moving to the right, superimposed on any other wave moving to the left.

Next we should check that this solution is at least intuitively consistent with the interpretation of the wave equation given at the end of the previous section: that the acceleration is proportional to the curvature. Take the case of a single bump moving to the right. Consider the motion of a single point at position A on the string (Fig. 3.1) as the bump passes by. The point slowly starts to move in the positive y direction, accelerates for a while, slows down, reaches its maximum deflection, and then reverses this process. This is analogous to a floating cork bobbing up and down as an ocean wave passes by.

Next consider the curvature of the bump as it passes by. It begins by growing slightly positive, then grows more positive, reaches a peak, flattens out to zero curvature, goes negative, reaches a negative peak (when the peak of the bump passes by), and finally reverses the process to return to zero. This is perfectly coordinated with its acceleration, which shows that the point's motion is at least consistent with the wave equation. The same argument works with the bump moving to the left. But this argument is neither precise nor very convincing: It doesn't predict the speed of the wave motion, and it doesn't predict that the shape of the bump will be preserved precisely.

Those results fall in our laps when we differentiate; sometimes we forget how much power is wrapped up so succinctly in our mathematical notation.

Up to now we haven't constrained the string in any way. It is infinitely long, not tied down at any point. We've seen that such a string can move in very general ways — the superposition of any two waves whatsoever moving in opposite directions. In particular, there's nothing about the form of Eq. 3.2 that predicts any particular pitch or periodic vibration. For this we must tie down the string at a couple of points, like a guitar string.

4 Reflection from a fixed end

Suppose next we fix the string at $x = 0$, so that it can't move at that point. This means that if we let $x = 0$ in the general solution Eq. 3.2, the deflection y must be zero, which yields the condition:

$$y(0, t) = f(t) + g(t) = 0 \qquad (4.1)$$

This must be true for every value of t, from which it follows that

$$f(t) = -g(t) \qquad (4.2)$$

The general solution therefore becomes

$$y(x, t) = f(t - x/c) - f(t + x/c) \qquad (4.3)$$

It's obvious that this automatically becomes zero when $x = 0$ for every t.

Equation 4.3 has a very interesting physical interpretation, illustrated in Fig. 4.1. Suppose we start a wave in the positive x region of the string, traveling left. This can be represented by the deflection function $y = f(t + x/c)$. We already know that if the string is to be fixed at $x = 0$ this cannot describe the entire deflection of the string. In fact, in order for the point of the string fixed at $x = 0$ to remain stationary when the bump arrives, there must be a component $-f(t - x/c)$ traveling to the right, which arrives at the origin at just the right time to cancel out any possible deflection at $x = 0$. This wave keeps traveling to the right, and the net effect is for the original wave to be reflected from the fixed origin with a reversal in sign, as shown in Fig. 4.1

As you might imagine, reflection of waves is a very important and general phenomenon in the study of sound. Next we will see how it allows us to understand the vibration of a string fixed at two points, and later, the vibration of air in tubes like organ pipes.

5 Vibration of a string fixed at two points

Suppose now we consider a string tied down at the point $x = L$, as well as $x = 0$. Mathematically this condition means that the displacement y is zero at the point $x = L$. Substituting this in Eq. 4.3 we get

$$y(L, t) = f(t - L/c) - f(t + L/c) = 0 \qquad (5.1)$$

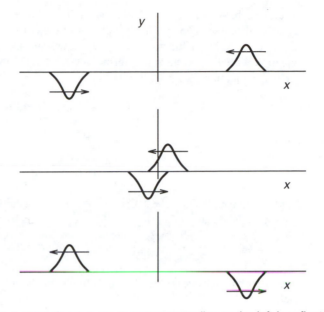

Fig. 4.1 A string fixed at $x = 0$; a wave traveling to the left is reflected with inversion. This is mathematically equivalent to its meeting a right-moving wave of opposite sign.

Since this is true for every value of t, it's permissible to add L/c to the arguments on both sides, yielding:

$$f(t) = f(t + 2L/c) \tag{5.2}$$

This tells us something quite significant: the displacement function $f(\cdot)$ is *periodic* with a period equal to $2L/c$ seconds. This period is the time that it takes for a wave to travel from one end of the string to the other and then back again — in other words, the round-trip time at velocity c.

This is a good time to mention a simple matter that sometimes causes confusion. If a waveform repeats itself every T seconds we say its *period* is T sec; its *frequency* is $f_0 = 1/T$ Hz (the reciprocal of its period). The unit *Hertz*, named after the German physicist Heinrich R. Hertz (1857–1894), can also be thought of as *cycles per sec*. Since there are 2π radians in a cycle, we also use *radian frequency* $\omega_0 = 2\pi f_0 = 2\pi/T$ radians per sec, which is convenient when we are discussing a sinusoid. For example, $\sin(2\pi f_0 t)$ repeats with the frequency f_0 Hz. So instead of writing the 2π all the time, we just use $\sin(\omega_0 t)$. In our case, $T = 2L/c$ sec, $f_0 = c/(2L)$ Hz, and $\omega_0 = \pi c/L$ radians per sec.

We are now going to make an educated guess at what a solution as a function of both t and x might be. We want to be sure that the deflection y vanishes at the endpoints $x = 0$ and $x = L$, but we know that the variation as a function of t is periodic with period $2L/c$, and has no such constraint. Therefore let's try a solution of the form

$$y(x, t) = e^{j\omega_0 t} Y(x) \tag{5.3}$$

where we have used $\omega_0 = \pi c/L$, the radian frequency corresponding to the period $2L/c$, as discussed above. This is a phasor of the correct frequency multiplied by some as yet undetermined function of x. We now want to see if we can satisfy the wave equation with a function of this form, so we calculate the left- and right-hand sides of the wave equation, Eq. 2.6:

$$\frac{\partial^2 y}{\partial t^2} = -\omega_0^2 e^{j\omega_0 t} Y(x) \tag{5.4}$$

and

$$c^2 \frac{\partial^2 y}{\partial x^2} = c^2 e^{j\omega_0 t} \frac{d^2 Y(x)}{dx^2} \tag{5.5}$$

For the wave equation to be satisfied, then, the right-hand sides of these last two equations must be equal:

$$-\omega_0^2 e^{j\omega_0 t} Y(x) = c^2 e^{j\omega_0 t} \frac{d^2 Y(x)}{dx^2} \tag{5.6}$$

The phasor factor due to the time variation cancels out, and the constant simplifies to yield

$$\frac{d^2 Y(x)}{dx^2} = -(\pi/L)^2 Y(x) \tag{5.7}$$

This should look familiar — it's exactly the same equation we used to describe the motion of a struck tuning-fork tine in Chapter 1, Eq. 2.2, and the result is simple harmonic motion. That is, the solution is

$$Y(x) = \sin(\pi x/L + \phi) \tag{5.8}$$

where the phase angle ϕ is yet to be determined.

Our guess has paid off. We've just verified that there is in fact a solution to the wave equation of the conjectured form, and that the function $Y(x)$, which determines the way the maximum deflection amplitude depends on x, is sinusoidal. But we still need to determine the angle ϕ, which establishes how the sinusoid is shifted relative to the beginning and end of the string.

Here's where we get to impose our condition that the deflection must be zero at the two ends of the string. It means that $Y(x)=0$ at $x=0$ and $x=L$, which implies from Eq. 5.8 that

$$\sin\phi = 0 \quad \text{and} \quad \sin(\pi + \phi) = 0 \tag{5.9}$$

This in turn implies that both ϕ and $\phi + \pi$ must be integer multiples of π. It doesn't matter which multiple of π we choose, so for simplicity we'll choose $\phi = 0$. Putting the two parts of $y(x, t)$ back together, we end up with

$$y(x, t) = e^{j\omega_0 t} \sin(\pi x/L) \tag{5.10}$$

A comment: Don't worry about this being a complex function. This didn't bother us in Chapter 1 and shouldn't bother us now. We'll just agree to take the real part if we want to have a real number for the displacement. For this reason we can also consider the solution obtained to be

$$y(x, t) = \cos(\omega_0 t)\sin(\pi x/L) \tag{5.11}$$

This solution has the following meaning: A point on the string at position x vibrates sinusoidally at the radian frequency $\omega_0 = \pi c/L$, with an amplitude that is greatest at the center of the string and decreases to zero at the end points. Note that this frequency varies inversely with the length of the string for a fixed wave velocity c. All else being equal, this predicts that the shorter the string, the higher the frequency of vibration, as we expect.

It's interesting to rewrite Eq. 5.10 in the form

$$\begin{aligned}
y(x, t) &= \frac{e^{j\omega_0 t}}{2j}[e^{j\pi x/L} - e^{-j\pi x/L}] \\
&= \frac{e^{j\omega_0(t+x/c)}}{2j} - \frac{e^{j\omega_0(t-x/c)}}{2j}
\end{aligned} \tag{5.12}$$

where we have used the identity $\sin\theta = [e^{j\theta} - e^{-j\theta}]/(2j)$, easily derived from Euler's equation. This verifies that the solution is in fact of the form used in Eq. 5.1, the difference between right- and left-traveling waves. When two traveling waves combine to produce a wave that appears stationary, we say that a *standing wave* is produced.

Next, notice that when we suggested a solution of the form used in Eq. 5.3, a phasor of frequency f_0, we could equally well have used a phasor of frequency $2f_0$, $3f_0$, or any integer multiple k of f_0. All these repeat every $1/f_0$ seconds; in fact, a phasor with frequency kf_0 repeats k times in that period. The same procedure as above then leads to solutions

$$y(x, t) = e^{jk\omega_0 t}\sin(k\pi x/L) \tag{5.13}$$

for any integer k.

The solutions in Eq. 5.13 represent different *modes* in which the string can vibrate. The solution for $k = 1$, as described above, vibrates with greatest amplitude at the center of the string and with smaller and smaller amplitude as we go from the center to the endpoints. This is shown as the first mode in Fig. 5.1. Consider next the second mode, for $k = 2$. The solution is

$$y(x, t) = e^{j2\omega_0 t}\sin(2\pi x/L) \tag{5.14}$$

Each point on the string vibrates twice as fast as a point on the mode-1 string. Furthermore, the center of the string doesn't move at all! The largest amplitudes can be found at the midpoints of the two halves, the points at $x = L/4$ and $x = 3L/4$. Similarly, the solution corresponding to any k has $k - 1$ places besides the end points that aren't moving, and k places of maximum amplitude. All of these $2k - 1$ points are equally spaced along the string at intervals $L/(2k)$. The first couple of higher modes are illustrated in Fig. 5.1 along with the first mode, which is called the *fundamental* mode of vibration. The points on the string that don't move are called *nodes*.

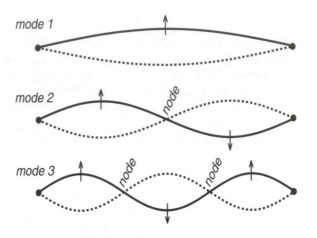

Fig. 5.1 The first three modes of a vibrating finite string. In mode 1 every point of the string moves in the same direction at any given time. In mode 2, the left half moves up when the right half moves down, and so forth. At the nodes, the string doesn't move at all.

We have now found a whole family of solutions to the wave equation, each member of which is zero at $x = 0$ and $x = L$, the ends of the string. We can now generate very general solutions by combining these in a simple way. To see how to do this we need two observations. First, notice that we can always multiply a solution to the wave equation by a constant factor without changing the fact that it's a solution. The constant factor will appear on both sides and cancel out. Second, if we have two solutions to the wave equation, the *sum* of the two solutions will also be a solution. This can be verified by substituting the sum of two solutions into the wave equation. The claim follows because the derivative of a sum is the sum of derivatives.

These two observations show that we can now find new solutions that are weighted sums of any modes we care to use. In general, therefore, we can use the grand combination

$$y(x, t) = \sum_{k=1}^{\infty} c_k e^{jk\omega_0 t} \sin(k\pi x/L) \tag{5.15}$$

where we have weighted the kth mode by the constant c_k. It turns out that this includes all the solutions that can possibly exist. To describe the precise pattern of vibration of any particular string, set into motion in any particular way, all we have to do is choose appropriate values for the constants c_k. If any mode is missing, the corresponding c_k is zero.

To sum up what we have learned about the vibrating finite string: Vibrations can exist only in a number of discrete modes, corresponding to integers k, one more than the number of nodes on the string (not counting the endpoints). The string vibrates at a frequency kf_0 in mode k, where the period $1/f_0$ is the round-trip time of a wave at the natural wave speed c determined by the tension and mass density of the string.

You should realize that the string cannot have a fundamental frequency of vibration until we specify a boundary condition at *two* points. In this case we prevent it from moving at two points. At that point we have defined a *length*, which defines a round-trip time, which in turn defines a frequency of vibration. In other words there is no way that an infinitely long string that is not tied down, or that is tied down at only one point, can vibrate periodically. Standing waves form on the string between the two enforced nodes. We will see the same general phenomenon later in this chapter in the case of a vibrating column of air.

We are on the verge of discovering some truly marvelous properties of series like the one in Eq. 5.15. But the French geometrician Jean Baptiste Joseph Fourier (1768–1830) beat us to it by a couple hundred years, and so they are called *Fourier series*. We will return to them at the end of this chapter and study them in more detail later on. They will give us great insight into the way sounds are composed of frequency components. Before that I want to discuss another common kind of physical system that is used to generate musical sounds — a column of air vibrating in a tube.

6 The vibrating column of air

We are all familiar with a vibrating column of air making a sound, in an organ or a clarinet, for example. We'll now derive the basic equation that governs this sort of vibration. But first a word of caution. The analysis of air movement we will carry out here is highly simplified, much more simplified than the corresponding analysis for a string. This is because the motion of a gas is often complicated by the formation of turbulence — eddies and curlicues of all sorts and sizes — that are very difficult to characterize with simple equations. These effects are often very important in the production of sound, so don't think that the present analysis is the final word.

That said, I hope you'll delight in the fact that the basic equation of motion for a column of air is the same wave equation we've been studying. This is despite the fact that sound production in a pipe and by a string differ in important ways. True, both kinds of oscillations occur because of the balance between elastic restoring forces and inertial forces that tend to make the restoring motion overshoot. But there the similarity ends; the motion of air involves longitudinal compression instead of lateral displacement.

To get a picture of how waves move in air, first remember that air is composed of molecules in constant motion. The higher the temperature the faster the average motion. At any temperature and at any point in space there is an average pressure, which we'll denote by p_0. Suppose we push suddenly on a plane in contact with the air, say with a loudspeaker. As shown in Fig. 6.1, the air in front of the plane becomes temporarily compressed because the molecules in front of the plane have been pushed. This region of compression then travels outward from the plane at a characteristic speed, the speed of sound. As the wavefront passes, molecules are suddenly pushed forward by the molecules behind them, and then return to their average position. This should all be visualized as motion relative to *average* position of the air molecules. In fact the air molecules are in constant random motion.

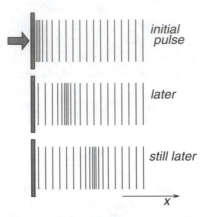

Fig. 6.1 Creation of a wavefront in air by sudden motion of a plane, and motion of the wavefront away from the plane.

The motion of the wavefront in air is analogous to the motion of a bump of lateral displacement along a stretched string, but the physics is different. In the first case the points on the string are moving up and down as the bump passes, at right angles to the direction of the wave motion. In the case of waves in air the individual molecules of air are moving randomly, and become locally displaced on the average as the wavefront passes, along the same axis as the wave motion. It is the deviation from a particle's average position that records the passage of the disturbance. We will measure this deviation from average position with the variable $\xi(x)$, where x is distance measured from the source of sound (see Fig. 6.1). I hope this isn't confusing; $\xi(x)$ is the local *deviation* from average position of a typical air molecule at position x. When there is no sound, $\xi(x) = 0$ for all x.

As I've pointed out in the cases of a tuning fork and stretched string, waves occur by a give and take between forces generated by elasticity and inertia. Our plan has the same general outline as before. We will first characterize the elasticity of air, which determines the force produced when we try to compress it. Then we will use Newton's second law to express the fact that air has inertia, and putting the two factors together will give us a differential equation of motion.

Visualize the air in a long cylindrical tube, sliced into very thin disks, as illustrated in Fig. 6.2.[†] A typical disk is bounded by two planes, the left plane at $x + \xi(x)$, and the right plane at $x + \Delta x + \xi(x + \Delta x)$. Remember that the variable $\xi(x)$ represents the *deviation* from the average position of the air molecules at position x. When no sound vibrations are present, $\xi(x) = 0$, and the thickness of the disk is Δx. When the air *is* vibrating, the thickness at any moment is

---

[†] The following derivation is classical, but I have leaned most on [Morse, 1948] (see the Notes at the end of this chapter).

$$\Delta x + \xi(x + \Delta x) - \xi(x) = \Delta x + \frac{\partial \xi}{\partial x} \Delta x \tag{6.1}$$

The last expression is the first-order approximation for the change in ξ with respect to x, which we are justified in using because Δx is infinitesimally small. We use the *partial* derivative because ξ is a function of both x and t, a fact we've ignored up to now to keep the notation simple.

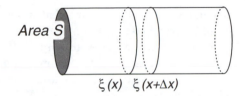

Fig. 6.2 Air in a long cylindrical tube, showing a typical infinitesimally thin slice (a disk).

We are next going to use the fact that the molecules in the space between the two faces of the slice always stay between the two faces. This is really just a way of saying that matter is conserved. If therefore the left face moves faster to the right than the right face, the air between becomes compressed; and if, conversely, the left face moves to the right more slowly than the right face, the air between becomes rarefied. Let ρ_0 be the density of the air at rest, with no vibration, and let the surface area of a face of the disk be S, as shown in Fig. 6.2. Then what we're saying is that the mass in the cylindrical slice is always the same. That is

$$\rho_0 S \Delta x = \rho S \Delta x (1 + \frac{\partial \xi}{\partial x}) \tag{6.2}$$

where ρ is the density of the slice at any moment. This equation allows us to express the ratio ρ/ρ_0 in terms of the derivative of ξ with respect to x. Specifically, the $S\Delta x$ cancels and we get

$$\frac{\rho}{\rho_0} = \frac{1}{1 + \partial \xi/\partial x} \tag{6.3}$$

The next step is to consider the pressure of the air at the faces of the slice. This will then allow us to find the difference in pressure at the two faces, and that will represent a force on the slice of air. There is first of all some steady ambient pressure p_0, which is immaterial. Only the changes in pressure matter, just as only the changes in position x of the molecules matter. Let us call the pressure change at any place and time p, so the total pressure is $p_0 + p$. Then the physical properties of gasses imply that the fractional change in pressure p/p_0 is proportional to the fractional change in density. That is,

$$\frac{p}{p_0} = \gamma \left[\frac{\rho - \rho_0}{\rho_0} \right] \tag{6.4}$$

where γ is some constant determined by the physical characteristics of the gas in question — air in this case — and is called a *coefficient of elasticity*. Intuitively this is simple enough: it says that a sudden compression of the slice by a certain fraction results in a proportionate increase in pressure. Actually, this relation is based on an assumption that the vibrations of the air are fast enough that the heat developed in a slice upon compression does not have enough time to flow away from the slice before it becomes decompressed again. This is called *adiabatic* compression and decompression.[†]

It's important to keep in mind that when sound propagates in air the relative changes of everything we're dealing with — pressure, density, position — are all very small. That is, we're dealing with very small excursions from equilibrium values. We're going to use this fact now to simplify Eq. 6.3, the expression for ρ/ρ_0 in terms of the spatial derivative of ξ. The right-hand side of Eq. 6.3 is of the form $1/(1 + z)$. Expand this in a power series

$$\frac{1}{1 + z} = 1 - z + z^2 - z^3 + \cdots \tag{6.5}$$

When z is very small we can ignore the terms beyond the linear, yielding the approximation

$$\frac{1}{1 + z} \approx 1 - z \tag{6.6}$$

Applying this to Eq. 6.3, we get

$$\frac{\rho}{\rho_0} = 1 - \frac{\partial \xi}{\partial x} \tag{6.7}$$

using the fact that $z = \partial \xi / \partial x$ is very small. Substituting this approximation in Eq. 6.4 yields

$$p = -\gamma p_0 \frac{\partial \xi}{\partial x} \tag{6.8}$$

Now we get to apply Newton's second law. Consider the difference in pressures on the left and right faces of a typical slice of air in the tube, as shown in Fig. 6.3. The pressure on the left face is $p_0 + p$; on the right face it's

$$p_0 + p + \frac{\partial p}{\partial x} \Delta x \tag{6.9}$$

where we have approximated the change in p across the slice to first order using the derivative, as we approximated the change in ξ to get Eq. 6.1. The net force on the slice is the difference between the two pressures times the surface area S, which is

$$-S \frac{\partial p}{\partial x} \Delta x \tag{6.10}$$

Notice that we subtracted the pressure on the right face from that on the left face, to yield net force in the positive x direction. If the pressure is increasing to the right, the

---

[†] We're going to leave the thermodynamics at that; for more discussion see [Morse, 1948], or [Lamb, 1925].

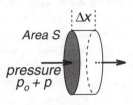

Fig. 6.3 A slice of air in a tube; the difference in pressure on the two faces results in a force on the mass of enclosed air.

force in Eq. 6.10 is negative (to the left), which makes sense because in this case there is more force on the right face than the left. Substitute p from Eq. 6.8 in Eq. 6.10, to get the net force

$$S\gamma p_0 \frac{\partial^2 \xi}{\partial x^2} \Delta x \qquad (6.11)$$

Finally, equate the mass of the slice times its acceleration to this net force. The mass is $(\rho_0 S\Delta x)$ and the acceleration is $(\partial^2 \xi / \partial t^2)$, so we get

$$\rho_0 S\Delta x \frac{\partial^2 \xi}{\partial t^2} = S\gamma p_0 \frac{\partial^2 \xi}{\partial x^2} \Delta x \qquad (6.12)$$

The volume of the slice $S\Delta x$ cancels out, and here we are again with the wave equation

$$\frac{\partial^2 \xi}{\partial t^2} = c^2 \frac{\partial^2 \xi}{\partial x^2} \qquad (6.13)$$

where the velocity of sound in air is

$$c = \sqrt{\gamma p_0 / \rho_0} \qquad (6.14)$$

Isn't it amazing that exactly the same equation governs both the vibration of a string and the vibration of air in a tube! But we are a long way from complete understanding. Why do they sound so different? There are many reasons, including the relative strength of the modes, and the very complicated things that happen to get the vibrations started in the first place. We'll get to some of those issues later, but next I want to discuss the most obvious and most easily understandable difference between standing waves on a string and in a tube.

7   Standing waves in a half-open tube

We saw earlier that the frequencies of the standing waves on a string are determined by its length. About the only thing we can do to set initial conditions for a string is to tie it down at two points, establishing its length. The mathematical condition corresponding to tying the string down at the point x is that its displacement $y(x)$ be zero. For a tube, this corresponds to the condition $\xi(x) = 0$, meaning that the displacement of air at the point x is forced to be zero. Closing off the tube with a solid wall means that air can't move there.

The finite tube that corresponds to a stretched finite string is closed at both ends. This is not a good way to make sound, at least not sound that we can hear. Usually, we excite the air at the closed end of a tube, with a vibrating reed, or lips, say, and leave the other end open. So we want to see what the standing waves are in a tube that is closed at one end and open at the other.

What is the mathematical condition that corresponds to the open end of a tube? The fact that the air at the open end communicates with the rest of the world means that it is free to expand or contract without feeling the effects of the tube. To a first approximation this means that deviations from the quiescent pressure p_0 cannot build up; in other words, the differential pressure $p = 0$. Equation 6.8 tells us that the differential pressure p is proportional to $\partial\xi/\partial x$, so the condition at the open end of the tube is

$$\frac{\partial\xi}{\partial x}\bigg|_{x=0} = 0 \tag{7.1}$$

instead of $\xi = 0$ at the closed end.

Let's see what this implies about the standing waves in a tube that is open at one end and closed at the other, a common situation. It really doesn't matter which way we orient the tube, so assume for convenience that the tube is open at $x = 0$ and closed at $x = L$. We know from the wave equation alone that the solution is of the form

$$\xi(x, t) = f(t - x/c) + g(t + x/c) \tag{7.2}$$

where $f(\cdot)$ is a right-moving wave that is of a completely arbitrary shape, and $g(\cdot)$ is a corresponding left-moving wave, also completely arbitrary. If we now enforce the condition in Eq. 7.1, for the open end of the tube, we get

$$f'(t) = g'(t) \tag{7.3}$$

This implies that the functions $f(\cdot)$ and $g(\cdot)$ differ by a constant, but constant differences in air pressure are immaterial to us and we are free to take $f(\cdot) = g(\cdot)$, which results in the total expression for the differential displacement

$$\xi(x, t) = f(t - x/c) + f(t + x/c) \tag{7.4}$$

This tells us that the reflection of a wave from the open end of the tube does *not* invert the wave, in contrast with reflection from the closed end, which *does*, being mathematically the same as reflection from a fixed point on a string. These reflections correspond to echoes, something we tend to take for granted. Why does sound bounce off the wall of a room or a canyon? It all follows from the beautifully concise wave equation.

To get standing waves at a definite frequency of oscillation, we need to impose a second condition, which is of course that the displacement ξ be zero at the point $x = L$. Substituting that condition in Eq. 7.4 yields

$$f(t - L/c) = -f(t + L/c) \tag{7.5}$$

This is true for every value of t, so we can add L/c to the argument of both sides, yielding

$$f(t) = -f(t + 2L/c) \tag{7.6}$$

We got almost the same condition in the case of a string tied down to zero at both ends, except the minus sign was missing. Now the function $f(\cdot)$ is periodic with period $4L/c$ instead of $2L/c$, a significant difference. We therefore now define the fundamental frequency ω_0 to correspond to this period, $2\pi c/(4L) = \pi c/(2L)$, and guess at the total solution

$$\xi(x, t) = e^{jk\omega_0 t}\,\Xi(x) \tag{7.7}$$

in analogy to Eq. 5.3. Notice that now we are considering the general case when the time oscillation has frequency $k\omega_0$, where k is any integer. When we considered the finite string we considered only the first mode, corresponding to $k = 1$, and the other modes were of the same form. Now, with the finite tube open at one end and closed at the other, it will be important to consider the more general case explicitly. Substituting in the wave equation Eq. 6.13 yields, again in analogy to the case of a string,

$$\frac{d^2\Xi(x)}{dx^2} = -\left(k\,\frac{\pi}{2L}\right)^2\Xi(x) \tag{7.8}$$

which tells us that the dependence on x is like that in a simple harmonic oscillator, of the form

$$\Xi(x) = \sin\left(\frac{k\pi x}{2L} + \phi\right) \tag{7.9}$$

where we have yet to determine the phase angle ϕ. To do this we again impose the conditions $\Xi'(0) = 0$ and $\Xi(L) = 0$, yielding

$$\cos\phi = 0 \quad \text{and} \quad \sin(\phi + k\pi/2) = 0 \tag{7.10}$$

A very interesting thing happens now. If the integer k is even, it is impossible for these two conditions to be satisfied simultaneously. To see this, rewrite $\cos\phi$ as $\sin(\phi + \pi/2)$, so the conditions become

$$\sin(\phi + \pi/2) = 0 \quad \text{and} \quad \sin(\phi + k\pi/2) = 0 \tag{7.11}$$

When k is even this means we are asking the sine function to be zero at two points that are an odd multiple of $\pi/2$ apart, which cannot happen. When k is odd, however, there is no problem. Therefore the solutions are all of the form

$$\xi(x, t) = e^{jk\omega_0 t}\cos\left(\frac{k\pi x}{2L}\right), \quad k = 1, 3, 5, \dots \tag{7.12}$$

Figure 7.1 shows the first three modes, corresponding to the values $k = 1$, 3, and 5. Compare this with Fig. 5.1, which shows what would happen if the tube were closed at both ends.

This has interesting implications about the way musical instruments work, and (just) begins to answer the earlier question of why strings sound much different from

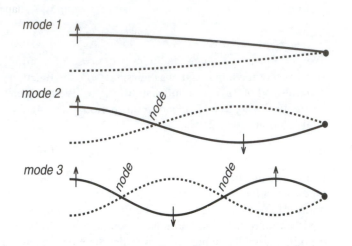

Fig. 7.1 The first three modes of a tube that is open at the left end and closed at the right.

wind instruments. In fact, wind instruments that depend on sound production by exciting a tube of air closed at one end and open at the other tend to be missing their even harmonics. More about that in the next section.

8   Fourier series

We've now studied two kinds of vibrating systems that are described by the wave equation, and derived a general mathematical form that describes the way they vibrate. Return to the vibrating finite string, and Eq. 5.15:

$$y(x,\, t) \; = \; \sum_{k=1}^{\infty} c_k e^{jk\omega_0 t} \sin(k\pi x/L) \qquad (8.1)$$

This is the mathematical way of saying the string's vibration can be broken down into an infinite number of discrete modes, with the kth mode having weight c_k. What determines the set of weights c_k for any particular sound? The answer is that they are determined by the particular way we set the string in motion.

Suppose we begin the string vibrating by holding it in a particular shape at time $t = 0$ and letting go. When you pluck a string, for example, you grab it with your finger, pull and let go. That would mean that the string's initial shape is a triangle. Let's imagine, though, that we can deform the string initially to any shape at all. At $t = 0$ Eq. 8.1 becomes

$$y(x,\, 0) \; = \; \sum_{k=1}^{\infty} c_k \sin(k\pi x/L) \qquad (8.2)$$

So far this may seem like an inconsequential thought-experiment, but the implications are far-reaching. This equation implies that any initial shape — that is, any function of x defined in the interval $[0, L]$ — can be expressed by this infinite series of weighted sine waves, provided we choose the weights c_k appropriately. I will leave the determination of the weights for a later chapter, but I want to emphasize now the intuitive content of this mathematical fact.

Next, I want to pull a switch that may be a bit startling, but mathematics is mathematics. We are free to think of Eq. 8.2 as describing an arbitrary function of *time* instead of space, say $f(t)$:

$$f(t) = \sum_{k=1}^{\infty} c_k \sin(k\pi t/T) \tag{8.3}$$

I have also replaced the length interval L by a time interval T. This can now represent any function of time in the interval $[0, T]$. The period of repetition is actually $2T$, because the sine waves that make up the series are necessarily antisymmetric about $t = 0$. That is, $f(t) = -f(-t)$ for all t. This determines $f(t)$ in the range $[-T, 0]$. When we return to the subject of Fourier series in earnest we will settle some obvious questions: How do we choose the coefficients c_k to get a particular shape? How do we represent functions that aren't antisymmetric?

The implications of Eq. 8.3 are familiar to musicians. The equation says that *any periodic waveform can be decomposed into a fundamental component* at a *fundamental frequency* (the $k = 1$ term), also called the *first harmonic*, and a series of higher *harmonics*, which have frequencies that are integer multiples of the first harmonic. This is illustrated in Fig. 8.1, which shows the measured spectrum of a clarinet note. To a first approximation, a clarinet produces sound by exciting a column of air in a tube that is closed at one end and open at the other. We get a bonus in this plot, because it tests the prediction that such a system does not generate even harmonics. In fact harmonics 1, 3, 5 and 7 are much stronger than harmonics 2, 4, and 6. (Note that the vertical scale is logarithmic, and is measured in dB.)[†] For example, the second harmonic is more than 30 dB weaker than the third. This pattern breaks down at the eighth harmonic and above. That's the difference between an ideal mathematical tube and a real live clarinet.

Speaking of deviations from a pattern, the sinusoidal components of sounds produced by musical instruments sometimes occur at frequencies different from exact integer harmonics of a fundamental frequency. When this happens the components are called *partials*. In some cases — bells, for example — the deviation of the frequencies of partials from integer multiples of a fundamental frequency can be quite large.

We'll return again and again to the idea that sound can be understood and analyzed mathematically by breaking it down into sinusoidal components. To a large extent our ears and brains understand sound this way — without any mathematics at all.

[†] Each 20 dB represents a factor of 10. More about this in the next chapter.

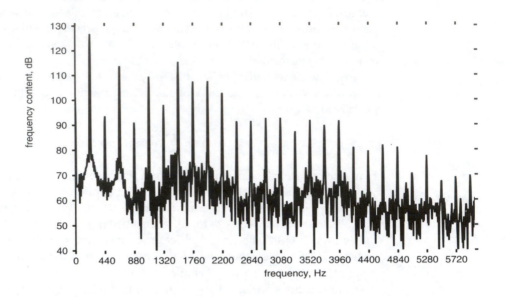

Fig 8.1 The frequency content (spectrum) of a note played on a clarinet. The pitch is A at 220 Hz, and the frequency axis shows multiples of 440 Hz. The first few even harmonics are very weak. If you're coming back to this from Chapter 10, or are an FFT aficionado, this plot was generated with a 4096-point FFT using a Hamming window, and the original sound was digitized at a sampling rate of 22,050 Hz.

Notes

I want to mention three famous books on sound, from which I've gotten most of the material in this chapter. I don't necessarily mean to recommend them as reading for you — they're old-fashioned and in some places downright stodgy. But each is a classic in its way.

First, there is the monumental and fascinating book by Lord Rayleigh,

> J. W. Strutt, Baron Rayleigh, *The Theory of Sound*, second edition, two volumes, reprinted by Dover, New York, N.Y., 1945. (First edition first published 1877, second edition revised and enlarged 1894.)

This book has the virtue of being written by a great genius who figured out a lot of the theory of sound for the first time. It's stylishly written and chock full of interesting detours, direct observations of experiments, and reports of his colleagues' work on subjects like difference tones, bird calls, and aeolian harps. If you run out of ideas for projects to work on, open randomly to one of its 984 pages.

Next comes a neat book that consolidates and simplifies much of the basic material in Lord Rayleigh,

[Lamb, 1925] H. Lamb, *The Dynamical Theory of Sound*, second edition, reprinted by Dover, New York, N.Y., 1960. (The second edition first published 1925.)

This book is a lot easier to read than Lord Rayleigh's.

Finally,

[Morse, 1948] P. M. Morse, *Vibration and Sound*, second edition, McGraw-Hill, New York, N.Y., 1948.

represents the progress made in the field through World War II. It's heavy reading, but I like the physical point of view.

Problems

1. The period of vibration of a stretched string is predicted to be $2L/c$ by Eq. 5.2. The period (reciprocal of the frequency) and the length are relatively easy to measure. This enables us to determine the wave velocity c on a stretched string. Do this for real stretched strings of your choosing, say your guitar strings, or piano strings. Compare the resulting velocities c with the speed of sound in air. Do you expect the velocities to be greater than or less than the speed of sound in air?

2. When the tension in a string is increased, does the velocity of sound along the string increase or decrease?

3. The velocity of sound in air is predicted by Eq. 6.14 to be $\sqrt{\gamma p_0/\rho_0}$. The quantity in this expression most difficult to measure directly is γ, the coefficient of elasticity. The velocity itself, the pressure at sea level, and the density are all known by direct measurement. Look them up and see what value they yield for γ.

4. Describe the form of solutions for vibration of air in a tube that is *open* at both ends, the expression analogous to Eq. 7.12.

5. Suppose you blow across one end of a soda straw, leaving the other end open. Then suppose that you block the other end with a finger. Predict what will happen to the pitch. Verify experimentally.

6. We've discussed the modes of vibration of strings and columns of air in pipes. Speculate about the vibration modes of a metal bar. Then verify your guesses by looking in one of the books given as reference.

7. Repeat for circular drum heads.

8. Suppose you pluck a guitar string, then put your finger in the center of the string, damping the motion of that spot. What do you think will happen to the spectrum of the sound? Verify experimentally.

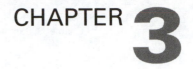

CHAPTER 3

Sampling and Quantizing

1 Sampling a phasor

I've spent a fair amount of time trying to convince you that the world is full of sinusoids, but up to now I haven't breathed a word about computers. If you want to use computers to deal with real signals, you need to represent these signals in digital form. How do we store *sound*, for example, in a computer? Let's begin with the process of *digitizing* real-world sounds, the process called *analog-to-digital (a-to-d) conversion*.

In most of this book I'll use sound waves like music and speech for examples. We're constantly surrounded by interesting sounds, and these waveforms are ideal for illustrating the basic ideas of signal processing. What's more, digital storage and digital processing of sounds have become part of everyday life.

Remember that the sound we hear travels as longitudinal waves of compression and rarefaction in the air, just like the standing waves in a tube. If we imagine a microphone diaphragm being pushed back and forth by an impinging wave front, we can represent the sound by a single real-valued function of time, say $x(t)$, which represents the displacement of the microphone's diaphragm from its resting position. That displacement is transformed into an electrical signal by the microphone. We now have two problems to deal with to get that signal into the computer — we need to discretize the real-valued time variable t, which process we call *sampling*; and we need to discretize the real-valued pressure variable $x(t)$, which process we call *quantizing*. An analog-to-digital converter performs both functions, producing a sequence of numbers representing successive samples of the sound pressure wave.

A digital-to-analog converter performs the reverse process, taking a sequence of numbers from the computer and producing a continuous waveform that can be converted to sound (pressure waves in the air) by a loudspeaker. As with analog-to-digital conversion we need to take into account the fact that both time and signal amplitude are discretized. We'll usually call the value of a signal a *sample value*, or *sample*, even if it isn't really a sample of an actual real-valued signal, but just a number we've come up with on the computer.

So there are two approximations involved in representing sound by a sequence of numbers in the computer; one due to sampling, the other due to quantizing. These approximations introduce errors, and if we are not careful, they can affect the quality of the sound in dramatic and sometimes unexpected ways. Let's begin with sampling and its effect on the frequency components of a sound.

Suppose we sample a simple sinusoidal signal. Analog-to-digital converters take samples at regularly spaced time intervals. Audio compact discs, for example, use samples that occur 44,100 times a second. The terminology is that the *sampling frequency*, or *sampling rate*, is 44.1 kHz, even if we're creating a sound signal from scratch on the computer. We'll reserve the symbol f_s for the sampling rate in Hz, ω_s for the sampling rate in radians per sec, and $T_s = 1/f_s$ for the interval between samples in seconds.

If the sampling rate is high compared to the frequency of the sinusoid, there is no problem. We get several samples to represent each cycle (period) of the sinusoid.

Next, suppose that we decrease the sampling rate, while keeping the frequency of the sinusoid constant. We get fewer and fewer samples per cycle. Eventually this causes a real problem. A concrete example is shown in Fig. 1.1, which shows 30 complete cycles of a sinusoid of frequency 330 Hz. Now suppose we sample it at a rate of 300 samples per sec. The resulting samples are shown as dots on the graph. If we had only the samples, we would think that the original signal is actually a sinusoid with a much lower frequency. What caused this disaster?

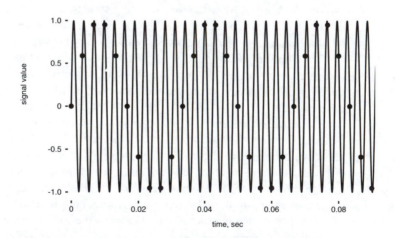

Fig. 1.1 Sampling a 330 Hz sinusoid at the rate of 300 Hz.

Intuitively the cause of the problem is obvious. We are taking samples every 1/300 sec, but the period of the sinusoid is 1/330 sec. The sinusoid therefore goes through more than one complete period between successive samples. What frequency do we think we're getting? To see this more easily, we'll return to our view of the sinusoid as the projection of a complex phasor.

Imagine a complex phasor rotating at a fixed frequency, and suppose that when we sample it, we paint a dot on the unit circle at the position of the phasor at the sample time. If we sample fast compared to the frequency of the phasor, the dots will be closely spaced, starting at the initial position of the phasor, and progressing around the circle, as shown in Fig. 1.2(a). We have an accurate representation of the phasor's frequency.

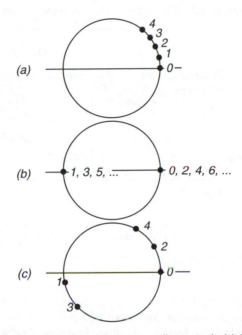

Fig. 1.2 Sampling a phasor. In (a) the sampling rate is high compared to the frequency of the phasor; in (b) the sampling rate is precisely half the frequency of the phasor; in (c) the sampling rate is slightly less than half the frequency of the phasor. In the last case the samples appear to move less than 180° in the clockwise (negative) direction.

Now suppose we gradually decrease the sampling rate. The dots become more and more widely spaced around the circle, until the situation shown in Fig. 1.2(b) is reached. Here the first sample is at the point $+1$ in the plane (the imaginary part is zero), the second sample is at the point -1, the third at $+1$, and so on. We know that the frequency of the sinusoid is now half the sampling rate, because we are taking two samples per revolution of the phasor. We are stretched to the limit, however.

Let's see what happens if we sample at an even slower rate, so that the frequency of the phasor is a bit *higher* than half the sampling rate. The result is shown in Fig. 1.2(c). The problem now is that this result is indistinguishable from the result we would have obtained if the frequency of the phasor were a bit *lower* than half the sampling rate. Each successive dot can be thought of as rotated a little less than π radians in the negative direction. As far as projections on the real axis are concerned, it doesn't matter which way the phasor is rotating. By sampling at less than twice the frequency of the phasor we have reached an erroneous conclusion about its frequency.

To summarize what we have learned so far, only frequencies below half the sampling rate will be accurately represented after sampling. This special frequency, half the sampling rate, is called the *Nyquist frequency*, after the American electrical engineer Harry Nyquist (1889–1976).

A little algebra will now give us a precise statement of which frequencies will be confounded with which. Write the original phasor with frequency ω_0 before sampling as

$$x(t) = e^{j\omega_0 t} \tag{1.1}$$

The nth sample at the sampling rate corresponding to the sampling interval T_s, which we'll denote by x_n, is

$$x_n = e^{j\omega_0 n T_s} \tag{1.2}$$

However, it is immaterial if we add any integer multiple of $j2\pi$ to the exponent of the complex exponential. Add $jnk2\pi$ to the exponent, where k is a completely arbitrary integer, positive, zero, or negative:

$$x_n = e^{j\omega_0 n T_s + jnk2\pi} \tag{1.3}$$

Rearrange this by factoring out T_s in the exponent, yielding

$$x_n = e^{jnT_s(\omega_0 + k2\pi/T_s)} \tag{1.4}$$

Equations 1.2 and 1.4 tell us that after sampling, a sinusoid with frequency ω_0 is equivalent to one with any frequency of the form $\omega_0 + k2\pi/T_s$. All the samples will be identical, and the two sampled signals will be indistinguishable from each other.

We now have derived a whole set of frequencies that can masquerade as one another. We call any one of these frequencies an *alias* of any other, and when one is confounded with another, we say *aliasing* has occurred. It is perhaps a little clearer if we replace $2\pi/T_s$ with ω_s, the sampling frequency in radians per sec. The aliases of the frequency ω_0 are then simply $\omega_0 + k\omega_s$, for all integers k.

Figure 1.3(a) shows the set of aliases corresponding to one particular positive frequency ω_0 a little below the Nyquist frequency. Aliases of it pop up a little below every multiple of the Nyquist frequency, being spaced ω_s apart by the argument above. This picture represents the algebraic version of our argument based on painting dots on the circle.

In real life, we sample real-valued sinusoids, not phasors, so we need to consider component phasors at the frequency $-\omega_0$ as well as $+\omega_0$. (This is because $\cos(\omega_0 t) = \frac{1}{2}e^{j\omega_0 t} + \frac{1}{2}e^{-j\omega_0 t}$.) The aliases of $-\omega_0$ are shown in Fig. 1.3(b). Because $-\omega_0$ is a little above the negative Nyquist frequency, its aliases pop up a little above every multiple of the Nyquist frequency. Finally, Fig. 1.3(c) shows the aliases

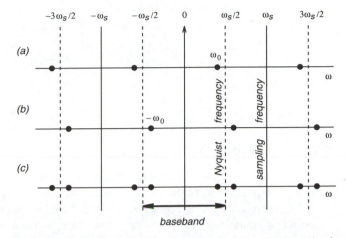

Fig. 1.3 (a) Aliases of the frequency ω_0 are shown as dots on the frequency axis; (b) aliases of the frequency $-\omega_0$; (c) aliases of both $+\omega_0$ and $-\omega_0$. Also shown are plus and minus the sampling frequency, ω_s; multiples of the Nyquist frequency, $\omega_s/2$; and the baseband between minus and plus the Nyquist frequency. If any of the dots in Part (c) appears in a signal, all the others are equivalent and can also be considered to be present.

of both $+\omega_0$ and $-\omega_0$. If a signal contains any one of these frequencies, all the others will be aliases of it.

We usually think of any frequency in a digital signal as lying between minus and plus the Nyquist frequency, and this range of frequencies is called the *baseband*. Any frequency outside this range is perfectly indistinguishable from its alias within this range. The frequency content of a digital signal in the baseband is sufficient to completely determine the signal. Of course we could pick any band of width ω_s radians per sec, but it's very natural to stick to the range we can hear. As a matter of fact though, an AM radio signal represents an audio signal in just this way by a band of width ω_s centered at the frequency of the radio transmitter.

Now observe a very important fact from Fig. 1.3(c) and the argument above. Every multiple of the Nyquist frequency acts like a mirror. For example, any frequency a distance Δf above $f_s/2$ will have an alias at a distance Δf below $f_s/2$. Perhaps this is obvious to you from the figures, but the algebra is also very simple. For every frequency ω_0 there is also the frequency $-\omega_0 + \omega_s$. The midpoint between them is the sum divided by two, or $\omega_s/2$, the Nyquist frequency. Thus, each frequency has an alias equally far from the Nyquist frequency but on the other side of it — on the other side of the mirror. For this reason the Nyquist frequency is often called the *folding* frequency because we can think of frequencies above Nyquist as being folded down below Nyquist.

I hope by now this seems simple. To summarize: If we're sampling at the rate ω_s, we can't distinguish the difference between any frequency ω_0 and ω_0 plus any multiple of the sampling rate itself. As far as the real signal generated by a rotating phasor, we also can't distinguish between the frequency ω_0 and $-\omega_0$. Therefore, all the following frequencies are aliases of each other:

$$\pm\omega_0 + k\omega_s, \quad \text{for all integers } k \tag{1.5}$$

2 Aliasing more complicated signals

We've seen the effect of aliasing on a single sinusoid. What happens if we sample a more complicated waveform — a square wave, for example? Here's where we begin to see the usefulness of Fourier series and the concept that all waveforms are composed of sinusoids. We introduced that concept in the previous chapter, where we saw that vibrating strings and columns of air very naturally produce sounds that can be broken into a fundamental tone and a series of overtones, harmonics at integer multiples of the fundamental frequency. We'll devote an entire subsequent chapter to the theory behind such periodic waveforms. But for now, just believe what I tell you about the Fourier series of a square wave.

We'll begin by considering sampling in the time domain. Figure 2.1 shows a square wave with fundamental frequency 700 Hz, sampled at the rate of 40,000 samples per sec. Mathematically, the waveform is defined by

$$f(t) = \begin{cases} +1 & \text{if } 0 < t < T/2 \\ -1 & \text{if } T/2 < t < T \end{cases} \tag{2.1}$$

which repeats with period T, $1/700$ sec. Since we're taking 40,000 samples per sec, there are $40,000/700 = 57^{1}/_7$ samples in each period. This means that different periods of the square wave will have different patterns of samples. In fact, from period to period, the first sample drifts to the left by $^1/_7$ of a sample period. Six out of seven periods contain 57 samples, and every seventh period contains eight samples. This averages out to the required $57^{1}/_7$ samples per period. That's the complete story in the time domain, but, as you can see, it doesn't shed much light on what sampling has done to the frequency content of the signal, and therefore tells us very little about what we can expect to hear. Thinking about this in the time domain is really not the answer. This is a good example of why it's often much better to look at things in the frequency rather than in the time domain.

As I warned above, I'm asking you to take the following Fourier series for our square wave on faith right now:

$$f(t) = \frac{4}{\pi}\left[\sin(\omega_0 t) + \frac{1}{3}\sin(3\omega_0 t) + \frac{1}{5}\sin(5\omega_0 t) + \cdots \right] \tag{2.2}$$

where ω_0 is the fundamental frequency, $2\pi/T$ radians per sec (or $1/T$ Hz). Although

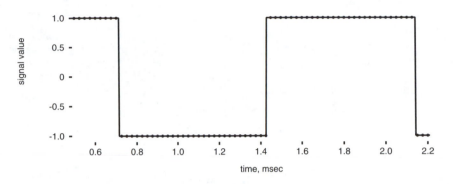

Fig. 2.1 A segment of a square wave, with samples shown. The fundamental frequency is 700 Hz, and the sampling rate is 40 kHz.

we have not yet described how we get the exact value of each coefficient, there are some important things about this series we can check.

First, the square wave we're using, as defined in Eq. 2.1, is arranged to be an *odd* function of t; that is, the waveform to the left of the point $t = 0$ is an upside-down version of the part to the right. Mathematically, $f(t) = -f(-t)$. Now sine waves have this property, but cosine waves don't. In fact, they're *even* functions, which means the part to the left of the point $t = 0$ is a rightside-up version of the part to the right. That is, $f(t) = f(-t)$. It is therefore entirely reasonable that the Fourier series be composed only of sine waves. Intuitively, the evenness of any cosine component would mess up the oddness of the sum, and it couldn't be fixed up with more sine waves.

Next, observe that our square wave is odd about the center of each period. That is, if we shift the signal so that it's centered at the half-period point, $t = T/2$, the resulting signal is also odd. Now sine waves at the odd harmonics have this property, but the sine waves at even harmonics are even about the midperiod point. Again, this is consistent with the Fourier series I've given in Eq. 2.2, which contains only the odd-numbered sine components. This sort of plausibility argument is very useful in checking Fourier series. (See Problem 8 for another example.)

Notice also that the nth harmonic has magnitude proportional to $1/n$. The higher harmonics decay in size to zero, but not very quickly. In general, we'll use the term *spectrum* to describe the frequencies in a signal, and we'll say this square wave has a spectrum that "falls off" as $1/n$. Later we'll learn more about the significance of the spectrum fall-off rate.

The key point is that we've broken down the square wave into sinusoidal components, and we can apply what we learned in the previous section about aliasing to each component separately.

Let's do the arithmetic for this case, in which the sampling frequency is 40 kHz and the fundamental frequency is 700 Hz. Figure 2.2 shows all the harmonics of the square wave as round dots. Harmonic numbers 1, 3, 5, . . . , 27 extend up to the

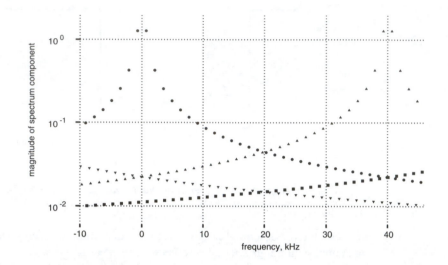

Fig. 2.2 Aliasing for a square wave. The sampling frequency is 40 kHz, and therefore the Nyquist frequency is 20 kHz. The round dots are the original components of the square wave, odd harmonics of the fundamental repetition rate of 700 Hz. The triangles and squares are aliases resulting from sampling. Note the logarithmic scale, which emphasizes the size of the aliased components.

Nyquist frequency of 20,000 Hz. The 27th harmonic is at 18,900 Hz, only 1100 Hz below the Nyquist. The next harmonic in the square wave, the 29th, occurs at 20,300 Hz, 300 Hz above the Nyquist. These frequencies also have their negative counterparts to the left.

From the results in Section 1, the harmonic at 20,300 Hz is folded down to 19,700 Hz. Similarly, the 31st harmonic, at 21,700 Hz, is folded down to 18,300 Hz. Also shown are aliases of more distant harmonics, which are of much lower amplitude. There are, in fact, an infinite number of harmonics aliased into the baseband. In Problem 9 I ask you to figure out where they all pile up.

While we're still on the subject of frequency content, and before we go on to quantizing effects, there's something very significant about the frequencies that are aliased into the baseband. In general, they are aliased to frequencies that are not integer multiples of the fundamental frequency of the square wave. (They can be by accident, if the sampling frequency is just right. See Problem 7.) Such components are called *inharmonic*. And it isn't just that they deviate slightly from true harmonics — they are utterly unrelated to any pitch we might hear that is determined by the fundamental frequency. If I didn't know better, I'd say they sound "bad." But computer musicians have taught me that there's no sound that isn't useful and interesting in the right context. You should listen for yourself (see Problem 11).

3 Quantizing

Computers today have many bits per word, more than enough to represent samples of sound pressure waves so precisely that we would never hear the effect of the finite word-length. The conversion processes — sampling with an analog-to-digital converter, or playing a sound with a digital-to-analog converter — cause the problems. Even high-quality converters usually have no more than 16, or at most 21, bits, and we must learn to use those bits wisely.

Let's assume that the amplitude values of a signal will be represented with B bits. There are 2^B possible values that can be represented by a B-bit word. For the sake of simplicity just think of the possible signal values as being equally spaced between -2^{B-1} and $+2^{B-1}$, to represent both negative and positive values. (In practice these numbers are scaled by some constant to represent the actual values.) It's common today for digital audio equipment to use 16-bit converters, so let's think of the case $B = 16$ for concreteness. This means we have at our disposal any of 65,536 possible values to represent any given sample value. The problem is that sample values can be any real numbers, like the values of sinusoids, for example. The quantizing process therefore entails an unavoidable approximation process, and introduces errors. How large are those errors and what effects might they have on the sound we ultimately hear? We can get very useful estimates using some pretty simple calculations.

Suppose a particular signal value is y, a real number in the range we've agreed to work in, -2^{B-1} and $+2^{B-1}$. Usually, we quantize y by rounding it off to the nearest integer, or what is called *quantizing level*. If the actual signal varies all over the place, typically jumping more than a few quantizing levels from sample to sample, then the relative position of the actual sample value with respect to the nearest quantizing level will be random and uncorrelated from sample to sample. Furthermore, the magnitude of the error will never be more than ½, so we can reasonably assume that it is uniformly distributed between the values 0 and ½. A close-up of quantizing a part of a sinusoid is illustrated in Fig. 3.1.

We next want to calculate a measure of the average error. Of course we shouldn't calculate the average value of the error itself, because it is just as likely to be negative as positive, and its average value is zero. Usually we deal with this problem by computing the average value of the *square* of the error, and then taking the square root of that. We call that the *root-mean-square* (*rms*) value of a signal. That way either a positive or a negative error will add to the total measure. By averaging the square we also reflect the fact that the ear responds to the *power* of a signal, rather than to its amplitude.

So let's calculate the rms value of the quantizing error. We argued above that it's uniformly distributed between 0 and ½, so the average value of its square is

$$\frac{1}{\frac{1}{2}} \int_0^{\frac{1}{2}} x^2\,dx = 1/12 \tag{3.1}$$

and its rms value is $1/\sqrt{12} = 0.2887$.

What really matters is the rms error *relative* to the level of the signal; that is, we should normalize the rms quantizing error by the maximum value of the signal, 2^{B-1}, yielding the ratio

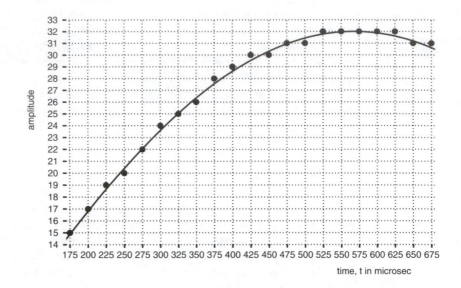

Fig. 3.1 Close-up of quantizing a sinusoid. The continuous curve is the original waveform, and the dots are samples, quantized to the nearest integer. In this particular example the frequency of the sinusoid is 440 Hz, and the sampling rate is 40 kHz, so that the samples are 25 µsec apart.

$$\frac{1/\sqrt{12}}{2^{B-1}} = \frac{1}{\sqrt{3}\,2^B} \tag{3.2}$$

The reciprocal of this ratio is what is called the *signal-to-noise* ratio, SNR = $\sqrt{3}\,2^B$. By dividing the rms noise by the maximum signal value we ensure that the signal-to-noise ratio as a measure of noise is unchanged if we just amplify the signal, which makes sense.

   This is a good time to talk about *decibels* (*dB*). It turns out that the ear responds to *ratios* of signal amplitude or power, rather than to arithmetic differences. For example, suppose we double the amplitude of a signal from 1 to 2. We perceive a certain increase in loudness. To get another increase in loudness that is perceived as roughly equal to the first increase, we need to double again to 4, rather than increase by 1 to 3. To convert these increases by ratios to arithmetic increases, it is convenient to use the logarithm of signal values, so the decibel measure of a ratio R of amplitudes is defined to be $20\log_{10} R$. An amplitude ratio of 2 is $20\log_{10}(2) = 6.02$ dB, and it is quite common to hear people refer to the effect of doubling the amplitude of a signal as increasing its level by 6 dB. In general we often describe multiplicative factors by speaking of adding or subtracting decibels. Because the power varies as the square of the amplitude, the decibel measure of a ratio R of powers is defined to be $10\log_{10} R$. As you might guess, a decibel is a tenth of a bel, and as you might guess further, the bel is named for Alexander Graham Bell (1847–1922).

We can now express the relative size of quantizing error in decibels:

$$\text{SNR} = 20\log_{10}(\sqrt{3}\,2^B) = 4.77 + 6.02B \text{ dB} \qquad (3.3)$$

With 16 bits we therefore get a signal-to-noise ratio of 101.1 dB.

By the way, the calculation of rms value is simple, but an even simpler estimate is good enough. We've argued that under reasonable circumstances the quantizing error is uniformly distributed between 0 and ½. Its *mean-absolute-value* is therefore ¼, compared with its rms value of $1/\sqrt{12}$. The ratio is $2/\sqrt{3}$, or only 1.25 dB.

Finally, you may notice the figure 96.3 dB instead of 101.1 dB in some places (like magazines). This uses the ratio of maximum amplitude (2^{15}) to maximum quantizing noise (½), giving $20\log_{10}(2^{16}) = 96.3$ dB. However, I would argue that we hear quantizing noise as continuous noise (like tape hiss), and so respond to its average power. That's why I use $1/\sqrt{12}$ instead of ½, and the ratio of ½ to $1/\sqrt{12}$ is $\sqrt{3}$, or precisely the 4.77 dB in Eq. 3.3.

This is all nitpicking, however. Signal-to-noise ratios in the range of 90 dB represent a very idealized situation compared to reality, as we'll see in the next section.

4   Dynamic range

The ear can handle an enormous range of sound pressure levels. Table 4.1 shows the power levels I in watts per m^2 corresponding to sounds from the threshold of audibility to the threshold of pain. The term *dynamic range* is used rather loosely to mean the ratio between the levels of the loudest and softest sounds we expect in a sound source or a system. So we might say, for example, that the ear has a dynamic range of 120 dB. That's a power range of a trillion to one, or an amplitude range of a million to one, and dealing with such a large possible range of amplitudes gives us problems.

	I, w/m^2	level, dB
Threshold of hearing	10^{-12}	0
ppp	10^{-8}	40
p	10^{-6}	60
f	10^{-4}	80
fff	10^{-2}	100
threshold of pain	1	120

Table 4.1 The range of sound intensity I (in units of power per unit area) from the threshold of hearing to the threshold of pain (from [Backus, 1977]). The level of 10^{-12} watts per m^2, which is approximately at the threshold of hearing at 1000 Hz, is conventionally defined to be 0 dB [Morse, 1948, Chapter 2].

Suppose, for example, that we plan to accommodate levels up to *fff* while recording an orchestra, and therefore represent the corresponding maximum amplitude levels with the values $\pm 2^{15}$, using a 16-bit analog-to-digital converter. A passage as loud as *fff* is rare, and most of the time the sound level will be much lower. A *ppp* passage, for example, will have amplitude levels a thousand times, or 60 dB, smaller. This means the effective dynamic range throughout the *ppp* passage is no longer about 100 dB, but more like 40 dB. Put another way, we reserve 3 decimal digits, or about 10 bits, for the blast, and that leaves only about 6 bits for the quiet passage. (As a check, Eq. 3.3 shows that the SNR corresponding to 6 bits is 40.9 dB.) This is the real reason we need converters with at least 16 bits, not the SNR of 100 dB.

The figures in Table 4.1 are also interesting because they give us some measure of the absolute power levels involved in sound signals. The total power produced by a symphony orchestra playing at full volume can be estimated roughly by assuming a level of *fff* at the surface of a quarter-sphere with radius 50 m. That comes to about 80 watts. Backus [1977] cites measurements reported in 1931, putting a large orchestra at 67 watts, which is quite consistent. Evidently making music is not a very efficient operation in terms of energy production. Talking is even more feeble in terms of power: speaking at an ordinary conversational level produces only about 10^{-5} watt [Morse, 1948, Chapter 2].

5 Remedies: Companding and prefiltering

In practice, quantizing and aliasing will always cause a certain amount of error. The ways people ameliorate the effects of these errors illustrate the two most basic signal-processing techniques, waveshaping and filtering. I'll discuss these next briefly, and then go on in the next three chapters to discuss digital filtering in much greater detail.

Obviously, we should try to use as many bits as possible when we quantize. If we get too many bits, we can always throw some away once we get the data inside the computer. But the more bits, the more expensive and slower the converter, up to the point where it becomes technologically impossible to do any better. Today it seems that 16 or 18 bits is a reasonable compromise between quality and expense, and we certainly want to make the most of those bits. The main problem, as I've emphasized in the previous section, is that we must allow for a very large dynamic range in sound, and must therefore deal with relatively low-amplitude signals a large proportion of the time.

A popular way to deal with this is to boost the low signal amplitudes relative to the high signal amplitudes before quantizing, and then compensate numerically after quantizing. To do this we pass the original analog signal through a nonlinear function shaped like the curve in Fig. 5.1, before the sampling and quantizing process. The idea is that the output signal, represented by points on the *y*-axis, is quantized at equally spaced points, and this corresponds to quantizing levels that are squeezed together at low input signal levels and spread apart at high input levels. In effect this gains accuracy in dealing with low-level signals, in return for a sacrifice in accuracy for high-level signals.

For example, if the curve in Fig. 5.1 has a slope of 2 at the origin, the input quantizing levels are spaced half as far apart as they would be without this preprocessing. This gives us the equivalent of another bit for low-level signals, and on the average the quantizing error in this range is halved. The signal-to-noise ratio is 6 dB higher at

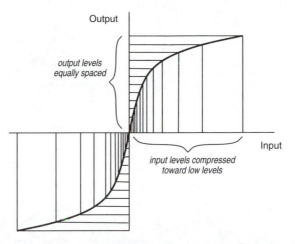

Fig. 5.1 Companding, or compressing and expanding. Sampling the output signal at equally spaced quantizing levels is equivalent to sampling the input signal at levels spaced closer together at low levels and farther apart at high levels.

low levels. But we don't get something for nothing; in order to accommodate signals with the same maximum amplitude as before, the curve must bend over to a slope less than 1, which means the equivalent of fewer bits for high-level signals. Of course we need to compensate for this intentional distortion when we receive the bits in the computer.

This general approach is called *companding*, which is short for "compressing and expanding." It is an example of processing a signal by using a nonlinear function of its value at any particular time. Since the output depends on the input signal value only at that particular time, we say the process has no memory, and the function in Fig. 5.1 is called an *instantaneous nonlinearity*. (See the Notes and Problem 12.)

Filtering, the way of dealing with aliasing error, is a fundamental and widely used technique in computer music, and in signal processing in general. The image conjured up by the word "filter" is quite appropriate — we will pass our original signal through the filter to remove some part of it, leaving the other parts unaffected. We want to remove the components above the Nyquist frequency so they won't be aliased down to frequencies in the usable range from 0 to the Nyquist. Figure 5.2 illustrates this idea. The original signal in general will have components above the Nyquist frequency, and, as we have seen, if we don't do anything about them, they will appear in the usable range below Nyquist after sampling. We therefore pass the signal through what we call a *lowpass* filter, one that affects the frequencies in the range $[-f_s/2, f_s/2]$ as little as possible, and eliminates all other frequencies as well as possible.

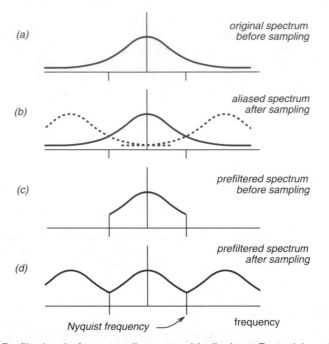

(a)   original spectrum
before sampling

(b)   aliased spectrum
after sampling

(c)   prefiltered spectrum
before sampling

(d)   prefiltered spectrum
after sampling

Nyquist frequency ⟶   frequency

Fig. 5.2 Prefiltering before sampling to avoid aliasing. Parts (a) and (b) show the signal spectrum before and after sampling with no prefiltering. Aliasing occurs; frequency components above the Nyquist frequency are aliased to frequencies below. Part (c) shows the signal spectrum after sampling if the original signal is prefiltered; the parts of the spectrum that would be aliased are removed before they can be folded down.

To take a concrete example, suppose we are going to sample a real-world signal, from a microphone, at the rate of 22,050 samples/sec. The signal may very well have frequency components above the Nyquist frequency, which is 11,025 Hz. We therefore want to pass the original signal through a lowpass filter that blocks all frequencies above 11,025 Hz but passes all those below that. It isn't possible to build such a filter perfectly, but we can come very close with careful design. Notice that this prefilter operates on the continuous signal, before sampling, and is not something we can implement on the computer. In other words, it is an *analog filter*, not a *digital filter*.

We'll return to the picture in Fig. 5.2 in Chapter 11, where we take up aliasing again in more depth. It turns out that aliasing can cause problems with digital-to-analog conversion as well as with analog-to-digital conversion, and the remedies are similar. This theory has a direct effect on our everyday life — audio and video compact discs work as well as they do because proper attention is paid to potential aliasing problems.

6 The shape of things to come

I hope by now it's natural for you to think of a signal as being composed of a sum of various frequency components. We concluded Chapter 2 with a mathematical argument to that effect, based on the fact that the arbitrary initial shape of a vibrating string can be expanded in a series of sine waves. This led to the Fourier series for any odd periodic signal with period $2T$:

$$f(t) = \sum_{k=1}^{\infty} c_k \sin(k\pi t/T) \tag{6.1}$$

As you might guess, the Fourier series that corresponds to the general case, when the periodic signal is not necessarily odd or even, is written in terms of a sum of sines and cosines. The more general form can also be written very neatly as the following sum of phasors:

$$f(t) = \sum_{k=-\infty}^{\infty} c_k e^{jk2\pi t/T} \tag{6.2}$$

This now represents any periodic signal $f(t)$ with period T sec as the sum of phasors with frequencies that are integer multiples of the frequency of repetition $2\pi/T$ Hz. Notice that we use phasors corresponding to both positive and negative frequencies. The spectrogram of the clarinet note shown in Fig. 8.1 of Chapter 2 is an experimentally measured version of such a Fourier series.

Let's go one step further. If we can represent any periodic signal as the sum of phasors with only those frequencies that are integer multiples of its frequency of repetition, how might we represent any signal whatsoever — even a nonperiodic one? Well, we need to incorporate phasors of *all* possible frequencies, not just the discrete set of integer multiples used in Eq. 6.2. To add these up we use an integral in place of the sum:

$$f(t) = \frac{1}{2\pi} \int_{-\infty}^{\infty} F(j\omega) e^{j\omega t} d\omega \tag{6.3}$$

Think of this as the grand sum of phasors of all possible frequencies, with the phasor of frequency ω present with weight $F(j\omega)$.[†]

The function $F(j\omega)$ tells us how much of each frequency we need to put in the integral to represent $f(t)$. It's what I've referred to loosely as the *spectrum* of a signal $f(t)$. That's how I used the term in the description of Fig. 5.2, for example, which shows what happens when a general signal is sampled and possibly aliased.

[†] The factor $1/2\pi$ in front of the integral is a mathematical bad penny. If you redefine the spectral weighting function $F(j\omega)$ to include it, it pops up somewhere else — in the formula for $F(j\omega)$ in terms of $f(t)$. Mathematicians sometimes define things so that there's a $1/(2\pi)^{1/2}$ in both places, for symmetry.

I've jumped ahead here because I wanted to encourage you to think more and more in terms of a signal's spectrum — the phasors that make it up. The next thing we'll study is filtering, which makes sense mostly in terms of its effect on a signal's spectrum.

Notes

The following book was written by a professor of physics for musicians and contains almost no mathematics. It gives some very nice physical intuition behind the operation of common musical instruments.

> [Backus, 1977] J. Backus, *The Acoustical Foundations of Music,* (second edition), W. W. Norton, New York, N.Y., 1977.

Backus has a small section at the end on computer music, and in the first edition he gives us a peek at the way things were at the beginning of time:

> One of the present problems in the use of a computer ... is the time lag — some hours to days — between the composer's instructions to the computer and the realization of the actual musical output as sound. ... Another problem with the computer is the expense: to produce a few minutes of music may require some ten times as much computer time at a cost of several hundred dollars per hour. [Backus, 1969 edition]

The following book has a similar slant, but is more comprehensive and up-to-date:

> A. H. Benade, *Fundamentals of Musical Acoustics*, Oxford University Press, New York, N.Y., 1976.

Benade discusses in detail the partials of chime and bell sounds, and emphasizes the distinction between partials and harmonics that I mentioned at the end of Chapter 2.

In digital-audio work you may run into a companding law called μ-law companding, which uses a particular form for the companding curve in Fig. 5.1 that is approximately linear at low levels, and logarithmic at high levels. For lots of details about quantizing, the definitive reference is

> N. S. Jayant and P. Noll, *Digital Coding of Waveforms: Principles and Applications to Speech and Video*, Prentice-Hall, Englewood Cliffs, N.J., 1984.

Instantaneous nonlinearities like the ones used for companding introduce new harmonics (*harmonic distortion*), and care must taken to reverse this effect by expanding after compressing. For this reason, unless we are companding, we usually avoid nonlinearities like the plague. But the effect can be exploited for the purposes of musical synthesis, and the resulting technique is called *waveshaping*. Curtis Roads attributes the origin and development of the idea to several people, starting with J.-C. Risset in 1969; R. Schaefer and C. Suen in 1970; and D. Arfib, J. Beauchamp, and M. LeBraun in 1979. See Roads's tutorial:

C. Roads, "A Tutorial on Non-Linear Distortion or Waveshaping Synthesis," *Computer Music Journal*, vol. 3, no. 2, pp. 29–34, 1979.

FM synthesis is another example of using instantaneous nonlinearities for musical synthesis. We'll discuss that technique in the final chapter.

Problems

1. What ratio of amplitudes is represented by one bel?

2. Aliasing can be observed in the world around you. Identify the source of the original signal and the sampling mechanism in the following situations:

(a) The hubcap of a car coming to a stop in a motion picture;

(b) A TV news anchor squirming while wearing a tweed jacket;

(c) A helicopter blade while the helicopter is starting up on a sunny day.

3. What frequency has been obtained by the sampling process illustrated in Fig. 1.1? The 330 Hz sinusoid is sampled at the rate of 300 Hz.

4. Sketch the first three terms of the Fourier series in Eq. 2.2 with pencil and paper, and add them up by eye. Check the symmetry properties of the sine waves that make it reasonable that this Fourier series adds up to the claimed square wave.

5. I claim in Section 2 that when a square wave with the repetition rate of 700 Hz is sampled at 40 kHz, the sampling pattern drifts to the left from period to period by $1/7$ of a period. To see this effect more clearly, do the case with pencil and paper when the repetition rate of the square wave is 30 kHz.

6. To what frequency in the baseband is the 79th harmonic of the square wave in Section 2 aliased?

7. A continuous periodic waveform with period P sec, and with all harmonics present, is sampled with sampling period T sec. Is it possible that for some choices of T and P the only frequencies that appear in the result are the ones in the original waveform below the Nyquist frequency? If the answer is yes, find conditions on T and P that ensure this happens; if the answer is no, prove it. Try to interpret your result in simple terms.

8. Suppose the time origin for the square wave in Fig. 2.1 were shifted to the right by $T/4$, a quarter of a period. Would the Fourier series contain sines? Cosines? Even harmonics? Odd harmonics? Repeat for a shift of $T/2$.

9. In the example of sampling a square wave in Section 2, the *sampled* waveform is periodic. What's the period? How is this periodicity reflected in the spectrum, which is illustrated in Fig. 2.2. What is the general relationship between the periods of the sampled and unsampled waveform?

10. (This should be easy after answering the previous question.) When does a waveform that is periodic become not periodic after sampling?

11. (Sound experiment) Generate the sound corresponding to a sampled square wave. Then generate the sound corresponding to sampling a square wave that has no harmonics above the Nyquist frequency, so there is no aliasing. Compare the sounds with and without aliasing. Can you find an interesting use for intentional aliasing?

12. Notice that

$$\cos(2\omega t) = 2\cos^2(\omega t) - 1$$

So if we let $x = \cos(\omega t)$, the signal

$$y = 2x^2 - 1$$

will be the second harmonic. This means we can get the second harmonic simply by passing x through an instantaneous nonlinearity.

(a) Work out the next case, $\cos(3\omega t)$ in terms of $\cos(\omega t)$.

(b) Prove the general fact that the result for every integer harmonic is always a polynomial. (The polynomials are called *Chebyshev* polynomials, after the Russian mathematician Paftnuty Lvovich Chebyshev [1821–1894].)

(c) Suppose we want to generate a signal with several cosine harmonics, each present with a predetermined amplitude. Describe a method for this using an appropriately designed instantaneous nonlinearity. (The method is called *waveshaping*; see the Notes.)

 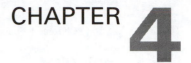

Feedforward Filters

1 Delaying a phasor

Filters work by combining delayed versions of signals. Guess what signal we'll delay to begin our study of filters? A phasor, of course.

If we delay the phasor $e^{j\omega t}$ by τ sec, we get

$$e^{j\omega(t-\tau)} = e^{-j\omega\tau}e^{j\omega t} \tag{1.1}$$

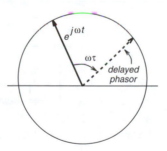

Fig. 1.1 The effect of delaying a phasor by τ sec. The dashed phasor is the delayed version of the original.

We see from this that a delay of τ sec multiplies the phasor by the complex factor $e^{-j\omega\tau}$, which does not depend on time t, but only on the amount of delay τ and the frequency ω. Such a factor can be written in the form $1\angle(-\omega\tau)$, and therefore rotates the original phasor by the angle $-\omega\tau$, while leaving its magnitude unchanged. This is illustrated in Fig. 1.1. Filters combine delayed versions of signals, and signals are made up out of phasors; so understanding the effect of this one operation on this one signal is the key to understanding everything there is to know about filters.

2 A simple filter

We're now going to build a very simple filter and analyze its effect on phasors of various frequencies. In fact, here's the simplest filter possible. Start with a signal x and add to it some constant a_1 times a delayed version of itself:

$$y_t = x_t + a_1 x_{t-\tau} \qquad (2.1)$$

Notice that we're using subscripts to denote the dependence on time. This will be especially convenient when we deal with digital signals in the computer because the time variable will just be an integer index in an array. Also, we'll usually try to reserve the symbols x and y for the input and output signals, respectively.

Sometimes it's helpful to draw a picture that represents the operations used to calculate the output of a filter from its input. Such a picture is shown in Fig. 2.1. The input signal enters from the left. A delayed version of it is obtained by tapping it at the junction indicated by the black dot and putting it into the box labeled with the delay τ. The addition of the signal to its delayed version is represented by the conjunction of arrows at the small circle labeled Σ for "summation." Finally, the output signal leaves on an arrow to the right. Figure 2.1 represents the flow of signals and is called a *signal flowgraph*. By the way, the convention is that signals flow from left to right as they progress through filters. When a delayed version of a signal is used later in the calculation, the branch representing the delay term goes from left to right, and can be thought of as feeding values "forward." For this reason I'll call filters that use only such branches *feedforward* filters. We won't get to feedback filters until the next chapter.

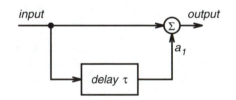

Fig. 2.1 Signal flowgraph of a simple feedforward filter.

Consider next what happens when x is a phasor at frequency ω, $e^{j\omega t}$. The right-hand side of Eq. 2.1 is then the sum of two phasors of the same frequency, which we know from Chapter 1 is also a phasor of that frequency. In other words, the output of the filter is a phasor of the same frequency as the input. Substituting for x we get

$$y_t = e^{j\omega t} + a_1 e^{j\omega(t-\tau)} \qquad (2.2)$$

This addition of two phasors is illustrated in Fig. 2.2; the second phasor is rotated by $-\omega\tau$ radians with respect to the first. In Chapter 1 we added phasors this way when we looked at beat frequencies, but this time there is the important difference that the second phasor isn't moving with respect to the first — it's just trailing behind by a fixed angle.

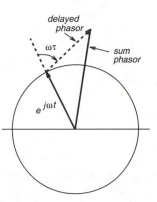

Fig. 2.2 Filtering: adding a phasor to a delayed version of itself. The angle between the original phasor and the delayed phasor is $-\omega\tau$.

Rewrite Eq. 2.2 by factoring out the phasor:

$$y_t = [1 + a_1 e^{-j\omega\tau}] e^{j\omega t} \tag{2.3}$$

This is a very significant equation. First, it shows that the output signal of the filter is a phasor at the frequency ω. Second, it says that the effect of the filter on the input phasor is to multiply it by the complex function in brackets on the right-hand side. Let's call the filter H, and denote that complex function by

$$H(\omega) = 1 + a_1 e^{-j\omega\tau} \tag{2.4}$$

which is called the filter's *frequency response*. Notice that $H(\omega)$ depends on the frequency ω and on the fixed parameter τ of the filter, *but not on the time t.* $H(\omega)$ tells us everything we need to know about what the filter does to the input phasor.

To see exactly how to use the information wrapped up in the frequency response $H(\omega)$, write it in polar form, as a magnitude at an angle:

$$H(\omega) = |H(\omega)| e^{j\theta(\omega)} \tag{2.5}$$

The magnitude $|H(\omega)|$ is called the *magnitude response* of the filter, and the angle $\theta(\omega)$ is called its *phase response*. Since the input phasor is multiplied by $H(\omega)$, this tells us that the filter *multiplies the size of the input phasor by the filter's magnitude response*, and *shifts its phase by the filter's phase response*.

Let's take a closer look at the magnitude response of our filter H. From Eq. 2.4:

$$|H(\omega)| = |1 + a_1 e^{-j\omega\tau}| \tag{2.6}$$

To put this in terms of real variables, just rewrite the magnitude as the square root of the sum of squares of the real and imaginary parts, yielding:

$$H(\omega) = \left| 1 + a_1^2 + 2a_1 \cos(\omega\tau) \right|^{1/2} \qquad (2.7)$$

We have now achieved what we set out to do. We can plot the magnitude response of the filter as a function of frequency ω, and this will tell us what effect the filter will have on a phasor of any particular frequency ω. Remember that we can think of any signal at all as being composed of a sum of phasors, so this will tell us a great deal about the effect of the filter on an arbitrary input signal.

For example, Fig. 2.3 shows the magnitude response when we choose the filter parameter $a_1 = 0.99$ and the delay $\tau = 167$ μsec. The first notch in the frequency response occurs when the cosine in Eq. 2.7 equals -1, which occurs when $\omega\tau = \pi$, or at the frequency $f = 1/(2\tau)$ Hz (remember that $\omega = 2\pi f$). This checks with the figure, which has a notch at 3 kHz. These notches occur whenever $\omega\tau$ is an odd multiple of π. For example, the next notch is at three times this frequency, around 9 kHz. It is easy to see that the peaks occur when $\omega\tau$ is an even multiple of π, starting with zero.

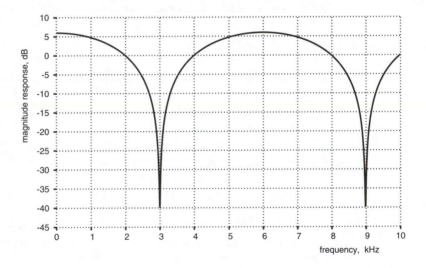

Fig. 2.3 Magnitude response (in dB) of a simple feedforward filter. The example shown has the filter parameter $a_1 = 0.99$, and the delay $\tau = 167$ μsec. Frequency is shown in kHz.

While we're at it, we should check the actual values of the magnitude response at the peaks and troughs. These are just $(1 + a_1)$ and $(1 - a_1)$, respectively, which translate into 1.99 and .01, or 5.977 dB and -40 dB.

The kind of filter in Eq. 2.1 is crude, but it does modify the spectrum of a signal in certain ways that might be useful — depending of course on what frequencies might be present in the signal to begin with. It reduces the presence of frequencies f_0, $3f_0$, $5f_0$, and so on, for the frequency $f_0 = 1/(2\tau)$, while leaving much of the remaining spectrum relatively unaffected. We'll see shortly, however, that it's just a toy compared to the really effective filters that are possible. Before we look at more complicated filters, though, we need to look at the limitations imposed by implementing filters on a computer.

3 Digital filters

We're going to concentrate entirely on filters implemented on a computer, which we'll call *digital filters*. The kinds of filters that can be implemented with components like acoustic delay lines, inductors, capacitors, and resistors, behave in similar ways, and are described by similar mathematics. But there are certain characteristics, both advantageous and restrictive, that are specific to digital filters. We'll begin with two crucial peculiarities.

In the previous section we assumed the kind of filter we studied can be implemented with any delay τ whatsoever. But when we use a computer we keep signals in arrays, and we are allowed to delay signals only an integer number of samples, because the time variable corresponds to the array index. On a computer, therefore, the delay τ must be an integer multiple of the sampling period T_s. That's one very important restriction.

There is another important restriction: In the digital world, frequencies above half the sampling frequency, the Nyquist frequency, don't really exist — we think of them as aliased to frequencies below the Nyquist frequency. If a phasor jumps more than π radians between sampling instants, we agree to think of it as jumping less than π radians per sample, thus giving us an unambiguous representation of frequencies less than the Nyquist, at the expense of not being able to represent any above the Nyquist. This means that the frequency plots of filter magnitude responses need not extend beyond the Nyquist frequency. It is usually convenient to normalize the frequency variable in such plots to the sampling rate, making the Nyquist frequency equal to 0.5. I will label the abscissas of such frequency-response plots "frequency, fractions of sampling rate."

I now will adjust notation slightly to simplify many equations in the rest of this book. Examine again the first equation in this chapter, which shows that delaying a phasor one sample, T_s sec, multiplies the phasor by $e^{-j\omega T_s}$. The frequency variable ω here has the units radians per sec. Therefore, ωT_s has the units radians per sample, and is the number of radians that the phasor turns between samples. We can always measure time in the digital domain in terms of the sample number, and we can always convert to actual time by multiplying by T_s sec. Therefore, in the digital world, we might as well think of ω as having the units radians per sample to begin with, and not write T_s with it all the time.

So from now on, in the digital domain, we'll measure the frequency ω in radians per sample. (In the continuous domain, we'll continue to use radians per sec.) The digital sampling frequency is then $\omega = 2\pi$ radians per sample (a full cycle between samples), and the Nyquist frequency is $\omega = \pi$ radians per sample (half a cycle between samples). The normalized frequency axis mentioned above, "frequency, fractions of sampling rate,'' can be thought of as measured in the units cycles per sample. To convert from this normalized frequency to actual frequency, multiply by the sampling rate.

A word about phasors. In the continuous world we write a phasor as $x_t = e^{j\omega t}$, where ω has the units radians per sec. In the digital world we'll write it in exactly the same way, remembering that ω is now measured in radians per sample, and interpreting t as the integer sample number. A delay of one sample in the digital domain therefore multiplies a phasor by $e^{-j\omega}$.

For a very simple example of a digital filter, suppose we use a delay of one sampling period in the filter equation Eq. 2.1:

$$y_t = x_t + a_1 x_{t-1} \tag{3.1}$$

where now the signals are indexed by the integer sample number t. When the digital signal x_t is the phasor $e^{j\omega t}$, the output phasor is

$$y_t = e^{j\omega t}\left[1 + a_1 e^{-j\omega}\right] \tag{3.2}$$

and the corresponding magnitude response of this digital filter is, as in Eqs. 2.6 and 2.7,

$$|H(\omega)| = |1 + a_1 e^{-j\omega}| = |1 + a_1^2 + 2a_1\cos\omega|^{1/2} \tag{3.3}$$

At the Nyquist frequency, ω is π radians per sample, and the cosine in Eq. 3.3 is equal to -1. When $a_1 > 0$ this means there is a dip at that point in the magnitude response. On the other hand, there is a relative peak at zero frequency, so this filter is *lowpass*, meaning it tends to pass low frequencies and reject high frequencies. Fig. 3.1 shows the frequency response of this filter for the value $a_1 = 0.99$. Because this is a digital filter, we need concern ourselves only with the frequencies below the Nyquist.

To summarize notation: In the continuous-time world, we'll use the continuous-time variable t sec and the frequency variable ω radians per sec; in the digital world we'll use the integer time variable t samples and the frequency variable ω radians per sample. The product ωt therefore measures radians per sec or per sample, depending on whether we are in the continuous or discrete domain. We've fussed a fair amount with notation in this section, but it will make life much simpler later on.

4 A big filter

I don't want to leave you with the impression that digital filters usually have only one or two terms. There's no reason we can't implement a filter with hundreds of terms; in fact this is done all the time. How can we possibly know how to select a few hundred coefficients so that the resulting digital filter has some desired, predetermined effect? This question is called the filter design problem. Fortunately, it's almost completely solved for feedforward digital filters. The mathematical problems involved were worked out in the 1960s and 1970s, and design packages are now widely

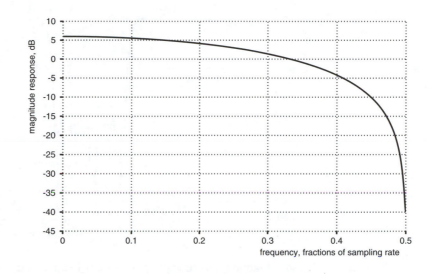

Fig. 3.1 Magnitude response (in dB) of a simple feedforward digital filter. The example shown has the filter parameter $a_1 = 0.99$ and a delay of one sampling period.

available. The particular filter used as an example in this section was designed using METEOR, a program based on linear programming [Steiglitz, et al., 1992].

Let's look at an example. Suppose we want to design a digital *bandstop* filter, which removes a particular range of frequencies, the *stopband*, but passes all others. The stopband is chosen to be the interval $[0.22, 0.32]$ in normalized frequency (fractions of the sampling rate). We require that the magnitude response be no more than 0.01 in the stopband, and within 0.01 of unity in the passbands. Figures 4.1 and 4.2 show the result of using METEOR for this design problem.

An interesting point comes up when we specify the passbands. Of course we'd like the passbands to extend right up to the very edges of the stopband, so, for example, the filter would reject the frequency 0.35999999 and pass the frequency 0.36. But this is asking too much. It is a great strain on a filter to make such a sharp distinction between frequencies so close together. The filter needs some slack in frequency to get from one value to another, so we need to allow what are called *transition bands*. The band between the normalized frequency 0.2 and 0.22 in this example is such a band. The narrower the transition bands, and the more exacting the amplitude specifications, the more terms we need in the filter to meet the specifications.

In the next section we'll start to develop a simple system for manipulating digital filters.

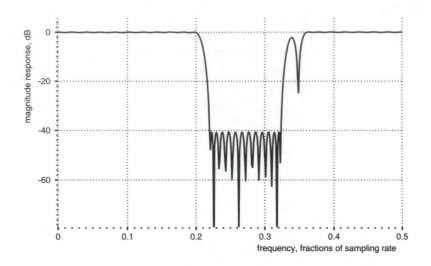

Fig. 4.1 Frequency response of a 99-term feedforward digital filter. The specifications are to pass frequencies in the interval [0, 0.2] and [0.36, 0.5], and to reject frequencies in [0.22, 0.32], all in fractions of the sampling frequency.

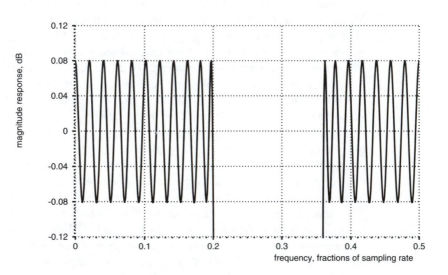

Fig. 4.2 Expanded vertical scale in the passbands of the previous figure.

5 Delay as an operator

To recapitulate, if the input to the following feedforward filter is the phasor $e^{j\omega t}$,

$$y_t = a_0 x_t + a_1 x_{t-1} \tag{5.1}$$

the output is also a phasor,

$$y_t = x_t\left[a_0 + a_1 e^{-j\omega}\right] \tag{5.2}$$

In general, with many delay terms, each term in Eq. 5.1 of the form $a_k x_{t-k}$ will result in a term of the form $a_k e^{-kj\omega}$ in Eq. 5.2.

Instead of writing $e^{j\omega}$ over and over, we introduce the symbol

$$z = e^{j\omega} \tag{5.3}$$

A *delay* of a phasor by k sampling intervals is then represented simply by multiplication by z^{-k}. Multiplication by z means the phasor is *advanced* one sampling interval, an operation that will be much less common than delay because it's more difficult or impossible to achieve in practical situations. (It's much harder to predict the future than to remember the past.) The simple feedforward filter in Eq. 5.1 is shown in the form of a flowgraph using this shorthand notation in Fig. 5.1.

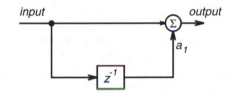

Fig. 5.1 Signal flowgraph of a simple feedforward digital filter.

Notation can have a profound effect on the way we think. Finding the right notation is often the key to making progress in a field. Just think of how much is wrapped up so concisely in Euler's formula or the wave equation, for example. A simple thing like using the symbol z^{-1} for delay is such an example. We're going to treat z^{-1} in two fundamentally different ways: as an operator (in this section) and as a complex variable (in the next). Both interpretations will be fruitful in advancing our understanding of digital filters.

An *operator* is a symbol that represents the application of an action on an object. For example, we can represent rotating this page $+90°$ by the operator ρ. If we represent the page by the symbol P, then we write ρP to represent the result of applying the operator ρ to P; that is, ρP represents the page actually rotated $+90°$. The operator ρ^{-1} is then the inverse operator, in this case rotation by $-90°$. The operator ρ^2 applied to a page turns it upside down; ρ^4 has no net effect — it's the identity operator — and so forth.

In the same way, let's use the symbol X to represent a signal with sample values x_t. Note carefully the distinction between x_t and X. The former represents the *value* of the signal at a particular time, and is a number; the latter represents the *entire* signal.

The signal delayed by one sampling interval is then represented by $z^{-1}X$. Here z^{-1} is an *operator*, which operates on the signal X. We can then rewrite the filter equation

$$y_t = a_0 x_t + a_1 x_{t-1} \tag{5.4}$$

as

$$Y = a_0 X + a_1 z^{-1} X = [a_0 + a_1 z^{-1}]X \tag{5.5}$$

Notice that I've slipped in another operator here. When I write $a_0 X$, for example, it represents the signal I get by multiplying every value of X by the constant a_0. It doesn't matter whether we multiply by a constant and then delay a signal, or first delay a signal and then multiply, so the order in which we write these operators is immaterial. In other words, the operator "multiply by a constant" *commutes* with the delay operator.

The notation of Eq. 5.5 is very suggestive. It tells us to interpret the expression in brackets as a single operator that represents the entire filter. We'll therefore rewrite Eq. 5.5 as

$$Y = \mathcal{H}(z)X \tag{5.6}$$

where

$$\mathcal{H}(z) = a_0 + a_1 z^{-1} \tag{5.7}$$

The operator $\mathcal{H}(z)$ will play a central role in helping us think about and manipulate filters; it's called the filter's *transfer function, $\mathcal{H}$.*

As a first example of how we can use transfer functions, consider what happens when we have two simple filters one after the other. This is called a *cascade connection*, and is shown in Fig. 5.2. The first filter produces the output signal W from the input signal X; the second produces the output Y from input W. Suppose the first filter has the transfer function

$$G(z) = a_0 + a_1 z^{-1} \tag{5.8}$$

and the second

$$\mathcal{H}(z) = b_0 + b_1 z^{-1} \tag{5.9}$$

The overall transfer function of the two filters combined can be written

$$Y = \mathcal{H}(z)W = \mathcal{H}(z)[G(z)X] \tag{5.10}$$

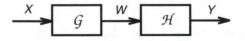

Fig. 5.2 The cascade connection of two digital filters.

The suggestive notation we've derived is now presenting us with a great temptation. Why not multiply the two operators together as if they were polynomials? If we do, we get

$$\mathcal{H}(z)\,G(z) = (a_0 + a_1 z^{-1})(b_0 + b_1 z^{-1})$$

$$= a_0 b_0 + (a_0 b_1 + a_1 b_0)z^{-1} + a_1 b_1 z^{-2} \tag{5.11}$$

Can we get away with this? The answer follows quickly from what we know about ordinary polynomials. We get away with this sort of thing in that case because of the distributive, associative, and commutative laws of algebra. I won't spell them all out here, but, for example, we use the distributive law when we write $\alpha(\beta + \gamma) = \alpha\beta + \alpha\gamma$. It's not hard to verify that the same laws hold for combining the operators in transfer functions: delays, additions, and multiplies-by-constants. For example, delaying the sum of two signals is completely equivalent to summing after the signals are delayed. We conclude, then, that yes, we are permitted to treat transfer functions the way we treat ordinary polynomials.

Multiplying the transfer functions in Eq. 5.11 shows that the cascade connection of the two filters in Fig. 5.2 is equivalent to the single three-term filter governed by the equation

$$y_t = a_0 b_0 x_t + (a_0 b_1 + a_1 b_0)x_{t-1} + a_1 b_1 x_{t-2} \tag{5.12}$$

This just begins to illustrate how useful transfer functions are. We got to this equivalent form of the cascade filter with hardly any effort at all.

Here's another example of how useful transfer functions are. Multiplication commutes; therefore filtering commutes. That means we get the same result if we filter first by H and then by G, because

$$G(z)\,\mathcal{H}(z) = \mathcal{H}(z)\,G(z) \tag{5.13}$$

Put yet another way, we can interchange the order of the boxes in Fig. 5.2. Is that obvious from the filter equations alone?

We now have some idea of how fruitful it is to interpret z^{-1} as the delay operator. It gives us a whole new way to represent the effect of a digital filter: as multiplication by a polynomial. Now I want to return to the interpretation of z as a complex variable.

6 The z-plane

We can gain some useful insight into how filters work by looking at the features of the transfer function in the complex z-plane. Let's go back to a simple digital filter like the one we used as an example earlier in the chapter:

$$y_t = x_t - a_1 x_{t-1} \tag{6.1}$$

The effect on a phasor is to multiply it by the complex function of ω

$$1 - a_1 e^{-j\omega} = 1 - a_1 z^{-1} \tag{6.2}$$

Remember that we introduced z as shorthand:

$$z = e^{j\omega} \tag{6.3}$$

If we now have any transfer function at all, say $\mathcal{H}(z)$, the corresponding frequency response is therefore

$$H(\omega) = \mathcal{H}(e^{j\omega}) \tag{6.4}$$

That is, to get the frequency response, we simply interpret the transfer function as a function of the complex variable z, and evaluate it for values of z on the unit circle. The range of values we're interested in runs from $\omega = 0$ to the Nyquist frequency $\omega = \pi$ radians per sample. This is the top half of the unit circle in the z-plane, as shown in Fig. 6.1.

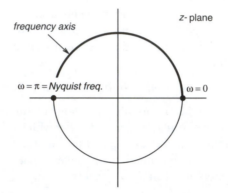

Fig. 6.1 The frequency axis in the z-plane. The top half of the unit circle corresponds to frequencies from 0 to the Nyquist frequency.

Let's take a closer look at the transfer function in our example. It's just

$$\mathcal{H}(z) = 1 - a_1 z^{-1} \tag{6.5}$$

Rewrite this as a ratio of polynomials, so we can see where the roots are:

$$\mathcal{H}(z) = \frac{z - a_1}{z} \tag{6.6}$$

There is a zero in the numerator at $z = a_1$, and a zero in the denominator at $z = 0$. That is, the transfer function becomes zero at a_1 and infinite at the origin.

The magnitude response is the magnitude of $\mathcal{H}(z)$ for z on the unit circle:

$$|H(\omega)| = \frac{|z - a_1|}{|z|} \quad \text{for } z = e^{j\omega} \tag{6.7}$$

The denominator is one, because z is on the unit circle. In fact, for feedforward filters the only possible zeros in the denominator occur at the origin, and these don't affect the magnitude response for the same reason — the magnitude of z on the unit circle is one. We can therefore rewrite Eq. 6.7 as

$$|H(\omega)| = |z - a_1| \quad \text{for } z = e^{j\omega} \tag{6.8}$$

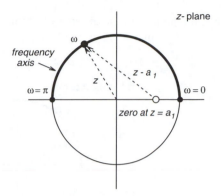

Fig. 6.2 Evaluating the magnitude response of a simple feedforward filter. The factor $|z - a_1|$ is the length of the vector from the zero at a_1 to the point on the unit circle corresponding to the frequency ω.

Figure 6.2 shows a geometric interpretation of this expression: it's the length of the vector from the zero at $z = a_1$ to the point on the unit circle representing the frequency ω at which we are evaluating the magnitude response.

This is an enlightening interpretation. Picture walking along the unit circle from 0 to the Nyquist frequency. When we are close to the zero, the length of this vector is small, and therefore so is the magnitude response at the frequency corresponding to our position on the unit circle. Conversely, when we are far from the zero, the magnitude response will be large. We can tell directly from Fig. 6.2 that a zero near $z = 1$ will result in a filter that passes high frequencies better than low — a highpass filter. On the other hand, a zero at $z = -1$ results in a lowpass filter.

The same idea works for more complicated feedforward filters. As an example, let's look at the following three-term filter:

$$y_t = x_t - x_{t-1} + x_{t-2} \tag{6.9}$$

I've picked simple coefficients, but what we're going to do will of course work for any feedforward filter. The transfer function is

$$\mathcal{H}(z) = 1 - z^{-1} + z^{-2} = z^{-2}(z^2 - z + 1) \tag{6.10}$$

I've factored out z^{-2}, two zeros in the denominator at the origin in the *z*-plane. We've just observed above that such factors don't affect the magnitude response. The second factor in Eq. 6.10 is a binomial, and can be factored further to exhibit its two zeros:

$$z^2 - z + 1 = (z - e^{j\pi/3})(z - e^{-j\pi/3}) \tag{6.11}$$

You see now that I've planned ahead; the zeros are precisely on the unit circle, at angles $\pm\pi/3$, or $f_N/3$, one-third the Nyquist frequency. The magnitude response is consequently

$$|H(\omega)| = |z - e^{j\pi/3}| \cdot |z - e^{-j\pi/3}| \quad \text{for } z = e^{j\omega} \tag{6.12}$$

which is the product of the distances from the two zeros to the point on the frequency axis, $z = e^{j\omega}$, as illustrated in Fig. 6.3.

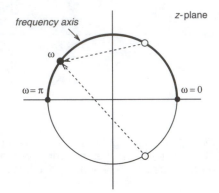

Fig. 6.3 Evaluating the magnitude response of the three-term, two-zero filter in Eq. 6.9. It's the product of the lengths of the two vectors from the complex zeros to the point on the unit circle corresponding to the frequency ω.

The magnitude response of this filter is plotted in Fig. 6.4. The zero at the point on the frequency axis corresponding to $f_N/3$ shows up as a sharp dip in this plot, a notch that theoretically goes down to $-\infty$ dB. If we apply any input signal to this filter, the frequency content of that signal around the frequency $f_N/3$ will be suppressed. Furthermore, if we apply a single, pure phasor precisely at the frequency $f_N/3$, this filter will not pass any of it, at least in the ideal situation where we have applied the input at a time infinitely far in the past. In practice, we have to turn the input on at some time, and this actually introduces frequencies other than one-third the Nyquist. More about this later.

We can develop some insight into the relationship between the coefficients and the magnitude response through the zero locations, but in a big filter we have to rely on high-powered design programs, like Parks-McClellan or METEOR (see the Notes), to do the proper delicate placement of zeros for a many-term example. Figure 6.5 shows the zeros of the 99-term filter used as an example in Section 4, with the magnitude response shown in Figs. 4.1 and 4.2. Notice the cluster of zeros in the stopband, as we would expect because the magnitude response is close to zero there. In fact there are some zeros precisely on the unit circle, and these correspond to frequencies where the magnitude response is precisely zero, as in the simple example of Figs. 6.3 and 6.4. Also notice all the other zeros; you wouldn't want to have to figure out where to put them by trial and error.

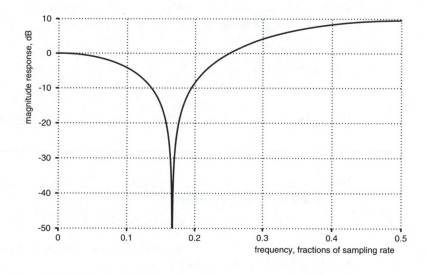

Fig. 6.4 Magnitude response of the three-term, two-zero filter in Eq. 6.9.

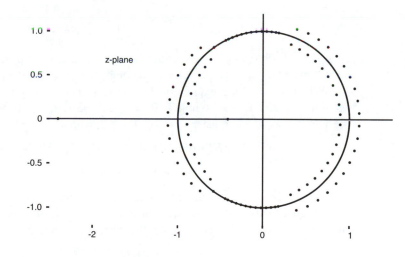

Fig. 6.5 The 98 zeros of the 99-term feedforward filter of Section 4.

7   Phase response

Up to now we've concentrated on the magnitude response of filters because that is most readily perceived. But filters also affect the phase of input signals. Consider again the simple kind of feedforward filter used as an example in Section 6:

$$y_t = x_t + a_1 x_{t-1} \tag{7.1}$$

with the frequency response

$$H(\omega) = 1 + a_1 e^{-j\omega} \tag{7.2}$$

Remember that for each frequency ω this has a magnitude and an angle; that is, it can be written

$$1 + a_1 e^{-j\omega} = |H(\omega)| e^{j\theta(\omega)} \tag{7.3}$$

If the input signal is the phasor $e^{j\omega t}$, the output phasor y is the product of that input phasor and the transfer function:

$$y_t = |H(\omega)| e^{j(\omega t + \theta(\omega))} \tag{7.4}$$

which means the output is shifted relative to the input by a phase angle $\theta(\omega)$, as well as having its size multiplied by $|H(\omega)|$.

We can evaluate this phase shift in a straightforward way using the arctangent function, as follows. Write Eq. 7.2 as $1 + a_1 \cos\omega - ja_1 \sin\omega$. Then

$$\theta(\omega) = \arctan\left[\frac{\mathit{Imag}\ H(\omega)}{\mathit{Real}\ H(\omega)}\right]$$

$$= \arctan\left[\frac{-a_1 \sin\omega}{1 + a_1 \cos\omega}\right] \tag{7.5}$$

It's also possible to get a geometric interpretation from a picture like Fig. 6.2. We won't examine phase response in great detail, except for one important point.

When $a_1 = 1$, the expression for phase response simplifies. We could use trigonometric identities to simplify Eq. 7.5, but the fastest way to see it in this simple case is to rewrite the transfer function, Eq. 7.3, with $a_1 = 1$:

$$H(\omega) = 1 + e^{-j\omega} = e^{-j\omega/2}[e^{j\omega/2} + e^{-j\omega/2}]$$

$$= e^{-j\omega/2} 2\cos(\omega/2) \tag{7.6}$$

The factor $2\cos(\omega/2)$ in this last equation is the magnitude response. The complex exponential factor is the phase response, and represents a phase angle of $-\omega/2$. If the input is the phasor $e^{j\omega t}$, the output of this filter is the phasor

$$2\cos(\omega/2) e^{j\omega(t - 1/2)} \tag{7.7}$$

This tells us that the effect of the filter on the phase of a phasor is to delay it precisely one-half sampling interval, *independent of its frequency*.

The preceding is a special case of the following important fact: If a feedforward filter has coefficients that are symmetric about their center (in this case the coefficients are $\{1, 1\}$), the phase response is *proportional to* ω, and that results in a fixed time delay. That is why one frequently hears the term "linear phase" mentioned as a desirable property of filters; it ensures that all frequency components of a signal are delayed an equal amount. In such cases we also say that the filter has no *phase distortion*. We'll see in the next chapter that it is not possible to achieve this property precisely with feedback filters.

8 Inverse comb filters

I will end this chapter by examining an important kind of feedforward filter, which I'll call an *inverse comb filter*. It is literally the inverse of a comb filter, which pops up in computer music all the time, and which we'll study in Chapter 6.

The inverse comb is very similar to the first feedforward filter we looked at, in Section 1. It consists of one simple feedforward loop, but instead of an arbitrary delay τ, we use some integer number of samples L. I'm going to choose the constant multiplier to be R^L, because we're going to take its Lth root very soon. The filter equation is

$$y_t = x_t - R^L x_{t-L} \tag{8.1}$$

and therefore its transfer function is

$$\mathcal{H}(z) = 1 - R^L z^{-L} \tag{8.2}$$

Let's take a minute to review a little complex algebra. Where are the roots of the following equation?

$$z^L = 1 \tag{8.3}$$

We know from the fundamental theorem of algebra that there are L of them. You may remember that they are equally spaced around the unit circle. That's kind of obvious when you think about what it means to raise a complex number to the Lth power: you simply raise its magnitude to the Lth power and multiply its angle by L. Therefore any point with magnitude 1 and angle of the form $k2\pi/L$, for integer k, will work. Thus the L roots of unity are

$$e^{jk2\pi/L}, \quad \text{for } k = 0, 1, \ldots, L-1 \tag{8.4}$$

It's now easy to see that the roots of Eq. 8.2,

$$z^L = R^L \tag{8.5}$$

are at the same angles, but at radius R instead of 1. The zeros for the case $L = 8$ are shown in Fig. 8.1.

The magnitude response corresponding to these zeros has dips at the frequency points near the zeros. The closer R is to 1, the deeper the dips. Figure 8.2 shows the magnitude response for the case $R = 0.999$. There are dips at multiples of one-eighth

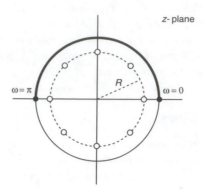

Fig. 8.1 The eight zeros of an inverse comb filter, equally spaced on a circle of radius R.

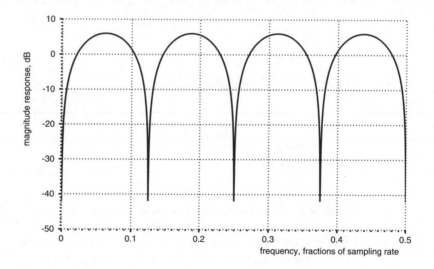

Fig. 8.2 The frequency response of the 8-zero inverse comb filter for the case $R = 0.999$.

the sampling rate. This magnitude response has the same shape as the one for the feedforward filter with arbitrary delay τ (see Fig. 2.3), except that now we can deal only with frequencies up to the Nyquist, and the dips are constrained to occur at integer multiples of an integer fraction of the sampling rate.

Now we're ready to study feedback filters. For an accurate preview of what's ahead, rotate Fig. 8.2 by π radians.

Notes 〜〜〜〜〜〜〜〜〜〜〜〜〜〜〜〜〜〜〜〜〜〜〜〜〜〜〜

My favorite program for designing feedforward filters is METEOR:

> [Steiglitz, et al., 1992] K. Steiglitz, T. W. Parks, and J. F. Kaiser, "METEOR: A Constraint-Based FIR Filter Design Program," *IEEE Trans. Signal Processing*, vol. 40, no. 8, pp. 1901–1909, August 1992.

FIR stands for Finite Impulse Response, and for practical purposes is synonymous with feedforward. METEOR allows the user to specify upper and lower bounds on the magnitude response of a feedforward filter, and then finds a filter with the fewest terms that satisfies the constraints. It uses linear programming; we'll describe it in more detail in Chapter 12.

The Parks-McClellan program is less flexible, but is easier to use and much faster:

> T. W. Parks and J. H. McClellan, "A Program for the Design of Linear Phase Finite Impulse Response Filters," *IEEE Trans. Audio Electroacoust.*, vol. AU-20, no. 3, Aug. 1972, pp. 195–199.

It's far and away the most widely used program for this purpose.

Problems 〜〜〜〜〜〜〜〜〜〜〜〜〜〜〜〜〜〜〜〜〜〜〜〜〜〜〜

1. Sketch the first few samples of a sinusoid at zero frequency, at the Nyquist frequency, and at half the Nyquist frequency. Is the answer to this question unique?

2. The simple filter used as an example in Section 2 suppresses frequencies at all odd multiples of 3 kHz. Design one that also uses Eq. 2.1 but that suppresses all *even* multiples of 3 kHz.

3. A feedforward digital filter is defined by the equation

$$y_t = x_t + x_{t-1} + x_{t-2} + \cdots + x_{t-99}$$

Derive a simple algebraic expression for its magnitude response. Find the frequencies at which its magnitude response has peaks and troughs.

4. Combine the defining equations for the two cascaded filters in Fig. 5.2 to prove directly the validity of the equivalent form in Eq. 5.12.

5. Write down and verify all the laws of algebra needed to justify the multiplication of transfer functions of feedforward filters as polynomials.

6. Prove that the transfer functions of feedforward filters with real coefficients can have complex zeros only in complex-conjugate pairs. What does this imply about the magnitude response? About phase response?

7. Design a feedforward filter that blocks an input phasor with frequency $f_N/6$, and also blocks zero frequency (DC bias). You should be able to do this with a four-term filter and without a lot of arithmetic. Plot its magnitude response. What is its peak magnitude response between 0 and $f_N/6$? Test it by filtering the signal

$$x(n) = 1 + \cos(\pi t/6)$$

where t is the integer sample number. Does the output ever become exactly zero?

8. Prove that the transfer function of a feedforward filter with coefficients symmetric about their center has linear phase. What is the resultant delay when the filter has N terms?

9. Plot the result of filtering the following signals with the linear-phase filter in Section 7, Eq. 7.1 with $a_1 = 1$:

(a) $x(t) = t$

(b) $x(t) = \sin(\pi t/100)$

(c) $x(t) = \begin{cases} +1 & \text{if } t \bmod 5 \text{ is even} \\ -1 & \text{if } t \bmod 5 \text{ is odd} \end{cases}$

In what sense does the filter introduce a delay of one-half a sampling interval?

10. Check the factorization of the polynomial $z^2 - z + 1$ in Eq. 6.11.

11. What is the value of the magnitude response shown in Fig. 8.2, that of an 8-zero inverse comb filter, at its minima? At its maxima?

CHAPTER 5

Feedback Filters

1 Poles

We're now going to use feedback to build filters. Feedback makes life more interesting — and more dangerous.

Suppose you were unhappy with the results you were getting with feedforward filters, and you were shopping around for something new to do. What do you have available in the filtering process that you haven't used? The only possibility seems to be past values of the *output* signal. Let's see what happens if we filter a signal x_t by adding a multiple of a past output value to it, instead of a past input:

$$y_t = x_t + a_1 y_{t-1} \tag{1.1}$$

This is the simplest example of a *feedback* filter, one in which we "feed back" past results in the computation of the next output value. Its signal flowgraph is shown in Fig. 1.1.

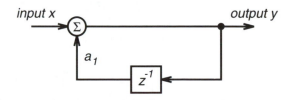

Fig. 1.1 Signal flowgraph of a one-pole feedback filter.

The first thing to notice is that if we turn off (set to zero) the input signal, the output signal can remain nonzero forever. For example, if $a_1 = 0.5$, and we put in the simple input signal that has the value one at $t = 0$ and is zero thereafter, the output signal takes on the values

$$y_t = 1, 0.5, 0.25, 0.125, 0.0625, \ldots \qquad (1.2)$$

This can never happen with a feedforward filter, because an output depends only on the inputs delayed by some maximum, finite amount, the greatest delay used in the filter equation. When the input is set to zero the output will become zero after that delay.

The next big difference between feedback and feedforward filters is that feedback filters can be *unstable*. To see this, suppose the coefficient in the filter of Eq. 1.1 is $a_1 = 2$, and we again supply an input signal that is one at $t = 0$ and zero thereafter. The output signal is

$$y_t = 1, 2, 4, 8, 16, \ldots \qquad (1.3)$$

and grows indefinitely. Of course this can never happen with a feedforward filter.

Let's take a look at the magnitude of the frequency response. The filter equation Eq. 1.1 can be written symbolically as

$$Y = X + a_1 z^{-1} Y \qquad (1.4)$$

There are a couple of ways to derive the transfer function from this. The simplest way to think about it is to rewrite this equation in a form that gives the input X in terms of the output Y:

$$X = Y - a_1 z^{-1} Y \qquad (1.5)$$

This is precisely the same form as a feedforward filter, but with input Y and output X. We know from the previous chapter that the transfer function of this feedforward filter is

$$1 - a_1 z^{-1} \qquad (1.6)$$

This transfer function evaluated for z on the unit circle at angle ω tells us the effect on the magnitude and phase angle of a phasor of frequency ω. Since this feedforward filter has an effect on a phasor that is the inverse of that of the feedback filter, the transfer function of the feedback filter must be

$$\mathcal{H}(z) = \frac{1}{1 - a_1 z^{-1}} \qquad (1.7)$$

and its magnitude response is therefore

$$|H(\omega)| = \frac{1}{|1 - a_1 e^{-j\omega}|} \qquad (1.8)$$

Another approach is to manipulate Eq. 1.4 with blind faith, solving for Y in terms of X. We have already justified multiplying, factoring, and permuting polynomial transfer functions as we do ordinary polynomials, but now we are proposing *dividing* both sides of the following equation by $1 - a_1 z^{-1}$:

$$Y[1 - a_1 z^{-1}] = X \qquad (1.9)$$

The result is identical to the one obtained above. We'll see a mathematically more rigorous way to justify this sort of operation later on, when we get to Fourier and z-transforms.

We now have a new kind of transfer function: one with a polynomial in the denominator. At a value of z where the denominator becomes zero, the transfer function becomes infinite. We call those points *poles* of the transfer function or filter. Figure 1.2 shows how we represent a pole in the z-plane by the symbol $\times$.

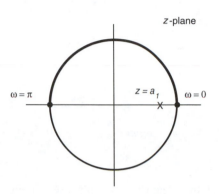

Fig. 1.2 Pole position in the z-plane for a single-pole feedback filter, Eq. 1.1.

We get the same kind of insight from the pole positions as we do from the zero positions. This time, however, we *divide* by the distance from a pole to a frequency point on the unit circle, as shown in Eq. 1.8. Visualize the magnitude frequency response as height above the z-plane, given by the function $|\mathcal{H}(z)|$, as we travel around the upper half of the unit circle from angle 0 to angle π radians. If we pass close to a pole, we will have a steep hill to climb and descend as we go by; the closer the pole the steeper the hill. If a pole is far away we will have to pass only a rolling foothill. If poles are at the origin, $z = 0$, they will have no effect on the magnitude response because we remain equidistant from them as we stay on the circle. This can also be seen from the fact that $|z^{-k}| = 1$ on the unit circle for any k. Figure 1.3 shows the magnitude frequency response of one-pole feedback filters for three different pole positions, illustrating what I just said.

2 Stability

Before we study the one-pole filter in more detail, and look at more complicated and interesting feedback filters, let's take a closer look at stability. The magnitude response is determined by the distance to poles, and is not affected by whether the response to a small input grows without bound, as in Eq. 1.3. But such an unstable filter would be quite useless in practice. In the case of the single-pole filter in Eq. 1.1, it's clear stability is ensured when and only when the pole is inside the unit circle; that is, when $|a_1| < 1$. But what about filters with many poles?

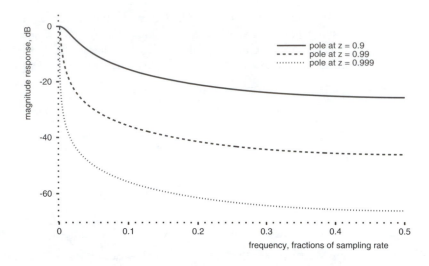

Fig. 1.3 Magnitude frequency response of single-pole feedback filters, Eq. 1.1, for pole positions a_1 = 0.9, 0.99, and 0.999. The plots are normalized so that the peak value is unity, or 0 dB.

The critical fact is that *any* pole outside the unit circle will cause a feedback filter to be unstable. Again, I want to postpone a proof until we get to z-transforms (Chapter 9), but will outline the general idea now. A feedback filter can have any number of feedback terms. Let's consider one with three, for example. It's defining equation looks like

$$y_t = x_t - b_1 y_{t-1} - b_2 y_{t-2} - b_3 y_{t-3} \tag{2.1}$$

The manipulation in the previous section leads to the transfer function

$$\mathcal{H}(z) = \frac{1}{1 + b_1 z^{-1} + b_2 z^{-2} + b_3 z^{-3}} \tag{2.2}$$

By now you see the pattern: The coefficients involving past output terms wind up in the denominator with their signs reversed. If we anticipate the sign reversal by using negative signs to begin with, as in Eq. 2.1, the signs in the denominator turn out to be positive. In this example the third-order denominator implies that there are three poles, and we can write the transfer function in factored form:

$$\mathcal{H}(z) = \frac{z^3}{(z - p_1)(z - p_2)(z - p_3)} \tag{2.3}$$

where the poles are at the denominator roots p_1, p_2, and p_3. Notice that we multiplied the numerator and denominator by z^3 so that the denominator is in the usual factored form of a polynomial. The factor of z^3 in the numerator means that there are three zeros at the origin. As we've seen, they won't affect the magnitude response or the stability. A pole can of course be complex, but because we assume the filter coefficients are real, if one is complex, its complex-conjugate mate must also be a root. That is, complex poles occur in complex-conjugate pairs.

Now it turns out that the transfer function in Eq. 2.3 can always be written in the following form:

$$\mathcal{H}(z) = \frac{A_1}{(1 - p_1 z^{-1})} + \frac{A_2}{(1 - p_2 z^{-1})} + \frac{A_3}{(1 - p_3 z^{-1})} \qquad (2.4)$$

You'll learn how to get the constants A_i in Chapter 9, but if you accept this for now you can see that it decomposes the original feedback filter into three one-pole feedback filters; the input is applied to all of them, and their outputs are added. This is called a *parallel connection* of filters, and is illustrated in Fig. 2.1.

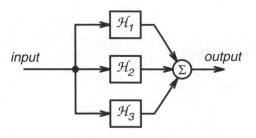

Fig. 2.1 Parallel connection of three one-pole filters $\mathcal{H}_1(z)$, $\mathcal{H}_2(z)$, and $\mathcal{H}_3(z)$; this corresponds to the decomposition of a three-term feedback filter.

Suppose we now test the original filter by applying a simple input signal, say an input that has the value one at $t = 0$ and zero otherwise. This particular test signal is called the *unit impulse*, and the output of a filter when a unit impulse is applied is called its *impulse response*. The output signal will be the sum of the outputs of each of the three one-pole filters. We know from Section 1 what each of these individual contributions is. If the ith one-pole filter has the transfer function $A_i/(1 - p_i z^{-1})$, the first output sample is A_i, the second is $A_i p_i$, and each successive output sample is multiplied by p_i. Ignoring the constant factor A_i, the output of each one-pole filter is the signal with samples

$$1, \ p_i, \ p_i^2, \ p_i^3, \dots \qquad (2.5)$$

Don't be bothered by the fact that the pole p_i may be complex; the one-pole feedback filter works the same way whether or not the pole is real, and if the pole *is* complex, its complex conjugate will appear in another one-pole filter of the total filter decomposition, and the outputs of the two will add to produce a real-valued total output.

How do we know that the unit impulse is a good signal with which to test stability? First, if the response to this particular signal is unstable, then certainly the filter can't be relied on to be stable in general. But what about the other way around? Might it not happen that when we apply the unit impulse, the filter is stable in the sense that the output does not grow exponentially, but some other input *will* cause the filter to be unstable? The answer is that this can't happen, unless of course we use an input that itself grows without bound. You can see this by the following intuitive argument (also see Problem 10). Any input signal can be thought of as a sum of delayed unit impulses, each weighted with a sample value. To be more precise, if we let δ_t denote the unit impulse signal, a general signal x can be written as

$$x_0\delta_t + x_1\delta_{t-1} + x_2\delta_{t-2} + \cdots \tag{2.6}$$

The term $x_0\delta_t$ represents the sample value at $t = 0$, the term $x_1\delta_{t-1}$ represents the sample at $t = 1$, and so on. The output when we apply this sum of impulses is what we would get if we applied each impulse individually, recorded the output, and added the results. From this we see that if the response to an impulse is stable, the response to *any* signal will be stable. We can summarize this argument by stating that a feedback filter is stable if and only if its impulse response is stable.

Returning now to the output of a one-pole filter when we apply the unit impulse signal δ_t, Eq. 2.5 shows that the critical issue is whether the magnitude of p_i is greater than one. If it is, the magnitude of the terms in Eq. 2.5 grows exponentially and the particular one-pole filter corresponding to p_i is unstable. The outputs of all the component one-pole filters are added, so if *any one* is unstable, the entire original filter is unstable. On the other hand, if all the component one-pole filters are stable, the original filter is stable.

One final detail before we wrap up this section. I've used the term "stability" loosely, and in particular I haven't spelled out how I want to regard the borderline case when a pole is precisely on the unit circle. If $|p_i| = 1$, the response of a one-pole filter to a unit impulse doesn't grow without bound. It doesn't decay to zero either, so in some sense the filter is neither stable nor unstable. I'll choose to be fussy and insist that a filter's impulse response must actually decay to zero for it to be called stable.

This line of reasoning works no matter how many poles a feedback filter has. We therefore have arrived at a very useful and general result: A feedback filter with poles p_i is stable *if and only if*

$$\boxed{|p_i| < 1 \text{ for all } i.} \tag{2.7}$$

3 Resonance and bandwidth

The bump in the magnitude response when we pass the part of the frequency axis near a pole is a *resonance*. It means the sizes of phasors at frequencies near that frequency are increased relative to the sizes of phasors at other frequencies. We are familiar with resonance in other areas. For example, soldiers marching across a bridge have to be careful not to march at a frequency near a resonance of the bridge, or the response to the repeated steps will grow until the bridge is in danger of collapse.

It's often important to know how sharp a particular resonance is given the position of the pole that causes it. We're going to derive that now. The measure of sharpness is illustrated in Fig. 3.1; it's defined to be the width of the magnitude response curve at

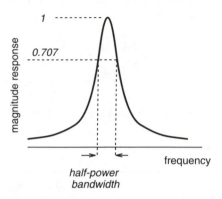

Fig. 3.1 The definition of bandwidth, the width of the magnitude response curve at the half-power points, the points where the squared amplitude is one-half the value at the peak.

the points where the curve has decreased to half the power it has at its peak, the so-called *half-power points*. Power is proportional to the square of the amplitude, so the half-power points correspond to amplitude values of $1/\sqrt{2}$ times the peak value (the -3 dB points). This measure is commonly called the *bandwidth* of the resonance.

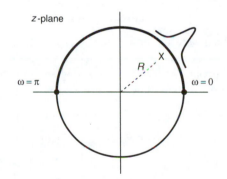

Fig. 3.2 A pole near the unit circle, causing a resonance.

A filter may have many poles, but we're going to assume that we're passing close to only one of them, and that all the rest are sufficiently far away that their effect on the magnitude response is very close to constant in the frequency range where the resonance occurs. Thus, we're going to consider the effect of one nearby pole, and ignore the effect of all the others. Figure 3.2 shows the situation, with a pole at a distance R from the origin in the z-plane, at some arbitrary angle.

Furthermore, we will assume that the pole in question is on the real axis, corresponding to a lowpass filter, with the peak in its magnitude response at zero frequency. This simplifies the algebra. When the pole is actually at some frequency other than zero, the effect on the magnitude response will be the same, but rotated by that angle. The inverse square of the magnitude response at frequency ϕ radians per sample[†] due to the one pole on the real axis at the point $z = R$ is, using Eq. 1.7 with $z = e^{j\phi}$ and $a_1 = R$,

$$\frac{1}{|H(\omega)|^2} = |e^{j\phi} - R|^2$$

$$= (\cos\phi - R)^2 + \sin^2\phi = 1 - 2R\cos\phi + R^2 \qquad (3.1)$$

At the center of the resonance, $\phi = 0$, this is equal to $(1 - R)^2$, just the squared distance from the pole to the zero-frequency point, $z = 1$. To find the half-power points, we simply look for the frequencies ϕ where the reciprocal power is twice this, so we solve

$$1 - 2R\cos\phi + R^2 = 2(1 - R)^2 \qquad (3.2)$$

for $\cos\phi$. The result is

$$\cos\phi = 2 - \frac{1}{2}(R + \frac{1}{R}) \qquad (3.3)$$

The bandwidth is simply twice the ϕ obtained here, the span from $-\phi$ to $+\phi$.

This solves the problem in theory, but we can also get a handy approximation for the bandwidth when R is close to unity, which is the most interesting situation — when the resonance is sharp. In that case, let $R = 1 - \epsilon$ and substitute in Eq. 3.3, yielding

$$\cos\phi \approx 1 - \epsilon^2/2 + \text{higher order terms in } \epsilon \qquad (3.4)$$

We used here the power series $1/R = 1/(1 - \epsilon) = 1 + \epsilon + \epsilon^2 + \cdots$. The first two terms on the right-hand side of Eq. 3.4 coincide with the beginning of the power series expansion for $\cos\epsilon$, so a good approximation for ϕ is $\phi \approx \epsilon$. Thus the bandwidth B in radians for R close to unity is

$$B = 2\phi \approx 2\epsilon = 2(1 - R) \quad \text{radians per sample} \qquad (3.5)$$

To get the pole radius R corresponding to a given bandwidth B, we rewrite this as

$$R \approx 1 - B/2 \qquad (3.6)$$

These last two equations tell us the relationship between pole radius and bandwidth for any resonant frequency, provided there are no other poles nearby.

To get a feeling for just how close R must be to one for a convincingly sharp resonance, let's take the example where the sampling rate is 44,100 Hz, and the bandwidth is 20 Hz. Equation 3.6 yields

$$R = 1 - \pi(20/44100) = 0.998575 \qquad (3.7)$$

Notice that we converted the bandwidth from Hz to radians per sample by dividing by the sampling rate in Hz and multiplying by 2π radians.

[†] Recall from Chapter 4, Section 3 that we've agreed to measure frequency in radians per sample, not radians per sec.

4 Resons

One-pole filters can resonate only at zero frequency or at the Nyquist frequency, depending on whether the pole is near the point $z = +1$ or $z = -1$. To get filters that can resonate at any desired frequency we need to use pairs of complex poles, and that's just what we're going to do in this section. The resulting filters are called *resonators*, or *resons*.

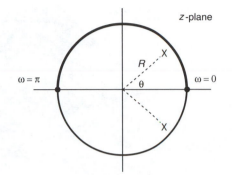

Fig. 4.1 Poles of a reson filter in the complex z-plane, at radius R and angle θ.

Figure 4.1 shows the position of the complex pole pair in the z-plane, at angles $\pm\theta$ and radius R. This corresponds to the transfer function

$$\mathcal{H}(z) = \frac{1}{(1 - Re^{j\theta}z^{-1})(1 - Re^{-j\theta}z^{-1})} \tag{4.1}$$

The two poles at $Re^{\pm j\theta}$ appear as the roots of the denominator factors. We've multiplied numerator and denominator by z^{-2} to keep things in terms of the delay operator z^{-1}; we'll always want to do this so that we can write the equation for the filter easily. All we need to do is multiply the denominator out as

$$\mathcal{H}(z) = \frac{1}{1 - (2R\cos\theta)z^{-1} + R^2 z^{-2}} \tag{4.2}$$

If we remember that this transfer function yields the output signal when multiplied by the input signal, interpreting z^{-1} as a delay operator (see Section 1), we can see immediately that this corresponds to the filter equation

$$y_t = x_t + (2R\cos\theta)y_{t-1} - R^2 y_{t-2} \tag{4.3}$$

We've already discussed the relationship between R and the bandwidth of the resonance. If the pole angle θ is not too near zero or π radians, the resonance at frequency θ is not affected much by the pole at $-\theta$, and the analysis in the previous section is valid. We can decide on the desired bandwidth and then determine R directly from Eq. 3.6.

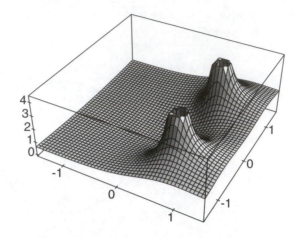

Fig. 4.2 Magnitude of the transfer function of a two-pole reson filter in the z-plane.

Figure 4.2 shows a perspective view of the magnitude of the transfer function of a reson filter above the z-plane; it illustrates how pole locations near the unit circle affect the magnitude response. Figure 4.3 shows the magnitude response for a family of three typical resons designed to have the same center frequency but different bandwidths.

Resons are very useful little building blocks, and we often want to design them to have particular bandwidths and center frequencies. We'll consider some details of their design in the next section.

5 Designing a reson filter

The next important thing to consider is the actual frequency of resonance. Many people assume that the peak of the magnitude response of a reson filter occurs precisely at θ, but this is not true. The second pole shifts the peak, and in certain cases the shift is significant (see Problem 3). The calculation of the true peak involves some messy algebra, but isn't difficult. I'm going to outline the derivation briefly here.

Our approach is simple-minded: We're going to evaluate the magnitude frequency response as a function of the frequency variable ϕ, and find a frequency ϕ at which the derivative is zero. To simplify matters we're going to consider not the magnitude of the frequency response, but its inverse-square. (The original frequency response has a maximum exactly when the inverse-square of the frequency response has a

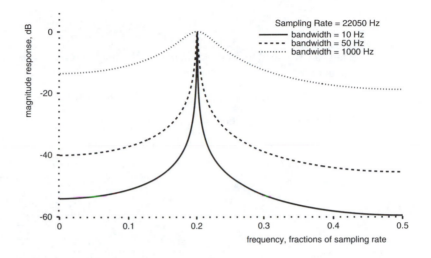

Fig. 4.3 Magnitude response of three resons, normalized so that the maximum gain is unity. The sampling rate is 22,050 Hz, all have the same center frequency of 4410 Hz, and the bandwidths are 10, 50, and 1000 Hz.

minimum.) Letting $z = e^{j\phi}$, the inverse-square of the magnitude response is, from Eq. 4.1,

$$\left| (e^{j\phi} - Re^{j\theta})(e^{j\phi} - Re^{-j\theta}) \right|^2 \tag{5.1}$$

This time it was convenient to multiply by the factor z^2 so that each factor is of the form $(z - \text{pole})$. The magnitude of z^2 is 1 on the unit circle, so we can do this without affecting the result. If we now use Euler's formula for the complex exponential, write the magnitude squared as the sum of real and imaginary parts squared, and simplify a bit, we get

$$(1 - R^2)^2 + 4R^2\cos^2\theta - 4R(R^2 + 1)\cos\theta\cos\phi + 4R^2\cos^2\phi \tag{5.2}$$

Differentiating with respect to ϕ (or $\cos\phi$) and setting the result to zero yields the value of ϕ where the peak actually occurs, which we'll call ψ:

$$\cos\psi = \frac{1 + R^2}{2R}\cos\theta \tag{5.3}$$

As expected, if R is very close to 1, ψ is very close to θ. This is easy to see because as R approaches 1, so does the factor $(1 + R^2)/(2R)$. Filter coefficients use the cosines of angles, rather than the angles themselves (as you can see from Eq. 4.3), and Eq. 5.3 is especially convenient because it allows us to find $\cos\theta$ given $\cos\psi$, or vice versa.

Finally, consider the scaling: We want to control the overall gain of a reson so that the amount it amplifies or attenuates a signal doesn't depend on its particular resonant frequency or bandwidth. One way to do this is to use a gain factor that makes the magnitude response unity at the resonant frequency ψ. This is easy; just substitute ψ from Eq. 5.3 into Eq. 5.2, the expression for the inverse-square of the magnitude response. The result, after some simplification, is

$$\frac{1}{|H(\psi)|^2} = (1 - R^2)^2 \sin^2\theta \tag{5.4}$$

Thus, all we need to do to normalize by the magnitude response at resonance is to use the gain factor

$$A_0 = (1 - R^2)\sin\theta \tag{5.5}$$

To summarize, here's how we usually design a reson:

(a) Choose the bandwidth B and the resonant frequency ψ.
(b) Calculate the pole radius R from the bandwidth B using Eq. 3.6:

$$R \approx 1 - B/2 \tag{5.6}$$

(c) Calculate the cosine of the pole angle θ using Eq. 5.3:

$$\cos\theta = \frac{2R}{1 + R^2}\cos\psi \tag{5.7}$$

(d) Calculate the gain factor A_0 using Eq. 5.5:

$$A_0 = (1 - R^2)\sin\theta \tag{5.8}$$

(e) Use the filter equation

$$y_t = A_0 x_t + (2R\cos\theta)y_{t-1} - R^2 y_{t-2} \tag{5.9}$$

Notice that the gain factor A_0 replaces unity in the numerator of the transfer function, Eq. 4.2, and hence appears as a multiplier of x_t in the filter equation.

6   Other incarnations of reson

The kind of system described mathematically by the two-pole reson filter is every-where. It is a more realistic version of the harmonic oscillator we started out with in the first chapter — more realistic because it takes the dissipation of energy into account. To understand what this means, let's look at the impulse response of reson, the response to the unit impulse input $x_t = \delta_t$.

The equation governing the generation of the impulse response y_t is, from Eq. 4.3,

$$y_t = \delta_t + (2R\cos\theta)y_{t-1} - R^2 y_{t-2} \tag{6.1}$$

The output values one sample old are multiplied by R, and output values two samples old are multiplied by R^2. This suggests that there is a factor R^t multiplying the solution, so we make the substitution

$$y_t = R^t w_t \tag{6.2}$$

and try to solve for w_t, which should be easier than solving for y_t. Substituting this in Eq. 6.1 we get

$$w_t = \delta_t + (2\cos\theta)w_{t-1} - w_{t-2} \qquad (6.3)$$

Notice that δ_t, the unit impulse, is zero for $t > 0$, which allows us to cancel the factor R^t for $t > 0$. When $t = 0$, $R^t = 1$, so there's no factor R to cancel. The signals are all zero for $t < 0$, so we can therefore cancel R for all t.

Equation 6.3 represents a two-pole resonator with poles exactly on the unit circle, at $e^{\pm j\theta}$. It's a good guess that the solution w_t is a phasor with frequency θ, but we don't know the phase angle. So let's postulate the solution

$$w_t = A\sin(\theta t) + B\cos(\theta t) \qquad (6.4)$$

where A and B are unknown constants, to be determined by the initial conditions of the filter.

Next, remember that we assume the filter output is zero for $t < 0$, so that we can compute the first couple of outputs by hand. This yields

$$w_0 = 1$$
$$w_1 = 2\cos\theta \qquad (6.5)$$

Substituting these values in Eq. 6.4 results in two conditions in two unknowns. The first tells us that $B = 1$, and the second tells us that $A = \cos\theta/\sin\theta$. A little rearrangement finally gives the impulse response in a neat form:

$$y_t = R^t\left[\frac{\sin(\theta(t+1))}{\sin\theta}\right] \qquad (6.6)$$

The response of reson has the following interpretation. The filter oscillates at the frequency θ, just as a harmonic oscillator does. If the pole radius R is one, putting the poles on the unit circle, that's all the filter does; it is in effect a sinusoid generator. The factor R^t means that the sinusoidal output is *damped* (assuming $R < 1$), decaying to zero. The smaller the pole radius R, the faster the output decays to zero. Figure 6.1 illustrates the impulse response for a typical center frequency and bandwidth.

This behavior is exactly analogous to what happens when a vibrating object loses energy to the surrounding world — and all eventually do. For example, a tuning fork eventually stops vibrating after being struck because it transmits energy to the air; otherwise we wouldn't hear it. This can be taken into account by adding a term that represents a restoring frictional force that is proportional to the velocity, so the equation for harmonic motion in Chapter 1 becomes

$$\frac{d^2x}{dt^2} = -(k/m)x - r\frac{dx}{dt} \qquad (6.7)$$

Notice the formal similarity between this differential equation and Eq. 4.3, which describes a two-pole reson. If we set the input to the reson filter to 0 and rearrange the equation, it takes the form

$$y_{t-2} = -ay_t - by_{t-1} \qquad (6.8)$$

This equation uses differences instead of derivatives, and it's an example of what is called a *difference equation*. In cases like these, with constant coefficients, the mathematical behavior of the two systems is the same.

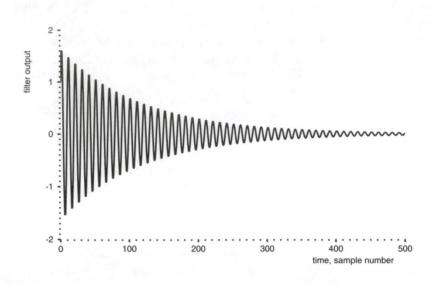

Fig. 6.1 Impulse response of a typical reson filter; the center frequency is 2205 Hz, the bandwidth 40 Hz, and the sampling rate 22,050 Hz.

Damped resonant systems are also old friends to electrical engineers. The RLC circuit shown in Fig. 6.2 shows what ham-radio operators would call a tank circuit. It resonates and can be used to tune receivers and transmitters. The smaller the resistance R, the narrower the bandwidth and the slower the damping. Building very sharp RLC filters is expensive because it's difficult to keep the resistance down. The analogous friction in digital filters is the roundoff noise that occurs because computer words store only a fixed number of bits.

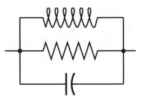

Fig. 6.2 A tuned RLC circuit, also known as a tank circuit; the physical counterpart of a two-pole reson.

7 Dropping in zeros: An improved reson

We can combine feedback and feedforward terms in the same filter without any problem. The result is a filter with both poles and zeros in its transfer function. The example I'll discuss in this section not only illustrates this, but also yields a useful improved version of reson.

The two-pole resonator discussed earlier in this chapter has a problem when the resonant frequency is close to zero or the Nyquist frequency, because in those cases the poles come close together and their peaks become partially merged. This is illustrated in Fig. 7.1, which shows pole locations corresponding to a low resonant frequency, and in Fig. 7.2, which shows the resulting magnitude response as the solid curve.

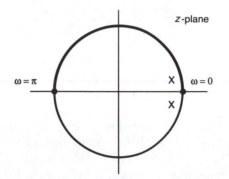

Fig. 7.1 Pole locations that result in a poor reson. The peaks of the two poles almost merge, and there is no real resonance at the intended low frequency.

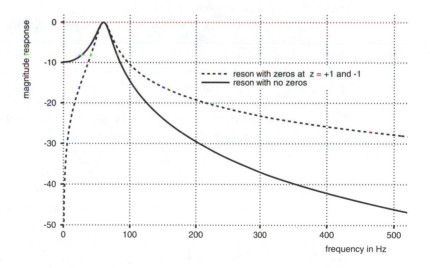

Fig. 7.2 Comparison of reson magnitude response with and without zeros. The case shown corresponds to a pair of poles near the zero-frequency point, as illustrated in Fig. 7.1. The specified center frequency is 60 Hz, the specified bandwidth is 20 Hz, and the sampling rate is 44,100 Hz.

The magnitude response is down only about 10 dB at zero frequency, and the "peak" at 60 Hz is hardly a peak at all.

One way to improve the shape of the reson is to put a zero at the point $z = 1$, which forces the frequency response to be zero at zero frequency. The same problem occurs at the Nyquist frequency, so we might as well put a zero there as well. This means multiplying the transfer function by the factor $1 - z^{-2}$, putting zeros at $z = \pm 1$. The new reson then has the transfer function

$$\mathcal{H}(z) = \frac{1 - z^{-2}}{1 - 2R\cos\theta z^{-1} + R^2 z^{-2}} \tag{7.1}$$

We'll call this new filter *reson_z*, to distinguish it from the original no-zero reson, which we'll call just *reson*. Figure 7.2 shows the magnitude response of reson_z along with the response of reson. The zero at zero frequency clearly introduces a deep notch there.

There is a price to pay for the improved selectivity of reson_z at low frequencies. The performance at frequencies much above the peak is not as good as reson. One way to explain this is to imagine yourself sitting on the unit circle in the z-plane at a point corresponding to high frequencies. In the case of reson_z, when you look back at the point $z = 1$ you see a cluster of two poles and one zero. One pole and one zero effectively cancel each other out, and from far away it looks as if there's only one pole at that point. In the case of reson, however, we see two poles at $z = 1$. The mountain created by the two poles is steeper than the one created by one pole; hence the magnitude response falls off that much more quickly for reson than for reson_z. A glance at Fig. 7.2 verifies this.

The idea of adding zeros to the two-pole reson was suggested by J. O. Smith and J. B. Angell [Smith and Angell, 1982], but not mainly to improve the shape of the resonance peak at very low (or very high) frequencies. The problem they were worried about comes up when we try to sweep the resonant frequency of reson by changing θ and keeping R constant, with the intent of moving the resonant frequency while keeping the bandwidth constant. It turns out that if you do this, the magnitude response at the peak can vary quite a lot. The new reson_z is much better behaved in this respect; I'll leave this for you to work on in the problems at the end of this chapter.

8   A powerful feedback filter

So far we've looked at feedback filters with only a couple of terms. How can we design more complicated and powerful feedback filters, comparable to the 99-term feedforward filter we saw in the previous chapter? There are two ways. First, we can try to adjust the pole and zero positions using iterative optimization methods. This turns out to be much more difficult than the corresponding problem for feedforward filters, essentially because the design criterion can no longer be formulated to be a linear function of the unknowns. The unknown parameters are in the denominator of the transfer function, and the resulting magnitude response is a much messier function than in the feedforward case.

The second approach to designing powerful feedback filters deals with only a few important, standard kinds of specifications, like lowpass, bandpass, and bandstop. Some very smart people worked on this in the 1930s, and left us remarkably convenient and useful "canned" filter designs. The most widely used, and in some ways the most impressive, of these designs is called the *elliptic* filter, and I'll show you an example in this section.

Poles have a lot more punch than zeros. Going to infinity at a point in the complex plane has a much more profound effect on the shape of the function around the point than merely taking on the value zero. This is related to the fact I've just mentioned, that feedback filters are hard to design to meet arbitrary specifications. But at the same time it means we need many fewer poles than zeros to achieve a desired effect. We'll see this in the example.

We'll look at a bandpass elliptic filter designed to pass frequencies in the range [0.22, 0.32] in fractions of the sampling rate and to reject frequencies outside this range. As usual, we can't hope to begin the bands of rejected frequencies exactly at the edges of the passband, but must leave a little room for the response to move between pass and stop, from zero to one or one to zero. The lower edge of the upper stopband is specified to be 0.36 times the sampling rate. The three numbers specified so far, giving the two edges of the passband and the lower edge of the upper stopband, determine the fourth, which is the upper edge of the lower stopband. The design algorithm for elliptic filters automatically determines the fourth frequency from the other three.

Only four pairs of complex poles and four pairs of complex zeros do the trick. Each pair of complex zeros and poles corresponds to a transfer function that looks like

$$\mathcal{H}(z) = \frac{1 + az^{-1} + bz^{-2}}{1 + cz^{-1} + dz^{-2}} \tag{8.1}$$

The filtering operation is usually broken down into stages like this, and the filtering is performed by passing the signal through successive stages. This is called a *cascade form* of *second-order* sections. This example, then, requires four such stages.

Figure 8.1 shows the positions of the poles and zeros. Notice that the poles are grouped around the passband, and the zeros occur exactly on the unit circle in the stopbands, to produce sharp notches there. Figure 8.2 shows the magnitude response of the feedback filter. There are many fewer ripples in the magnitude response than in the 99-term feedforward example given in the previous chapter because there are only eight zeros and eight poles.

Why would anyone ever use a feedforward filter with, say, 99 terms, when they could use a feedback filter with only 16 (four stages with four terms each)? One reason is the phase response. Feedforward filters can be designed with precisely linear phase, which means all frequencies are delayed equally. This amounts to having no phase distortion. Feedback filters, on the other hand, can have quite a bit of phase distortion. Another important reason is that feedforward filters are much easier to design to meet completely arbitrary specifications.

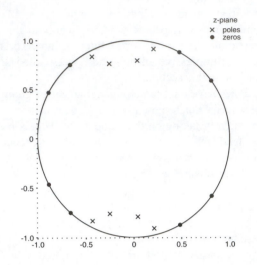

Fig. 8.1 The pole-zero pattern of the example elliptic passband filter.

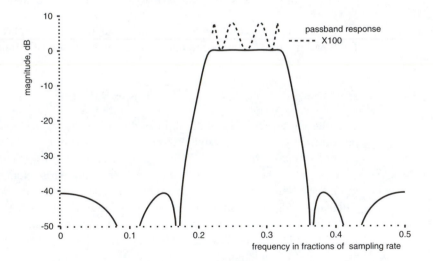

Fig. 8.2 Magnitude frequency response of the example elliptic passband filter. The passband is specified to be [0.22, 0.32] in fractions of the sampling rate, and the lower edge of the upper stopband is 0.36 in fractions of the sampling rate. The dashed curve in the passband shows the response magnified by 100. The specifications call for at most 0.083 dB ripple in the passband and at least 40 dB rejection in the stopband.

Notes

What I call feedback digital filters correspond almost always to what are called Infinite Impulse Response (IIR) filters in the literature. An IIR filter is defined to be a filter whose impulse response has an infinite number of nonzero terms. A two-pole reson, then, is an IIR filter as well as a feedback filter. But the terms are not exactly equivalent. Similarly, what I call feedforward digital filters usually correspond to what are called Finite Impulse Response (FIR) filters. To explore the subtle differences in definition, see Problem 9.

Note that the way I use the term, a feedback filter can have some feedforward terms, like the reson with added zeros in Section 7. What's important is whether or not there's feedback.

Adding zeros to reson to regulate peak gain was suggested in

> [Smith and Angell, 1982] J. O. Smith and J. B. Angell, "A Constant-Gain Digital Resonator Tuned by a Single Coefficient," *Computer Music Journal*, vol. 6, no. 4, pp. 36–40, Winter 1982.

The minutiae that are the subject of Problem 8 come from

> K. Steiglitz, "A Note on Constant-Gain Digital Resonators," *Computer Music Journal*, vol. 18, no. 4, pp. 8–10, Winter 1994.

Problems

1. The following approximation is sometimes given for the pole radius R of a reson in terms of the bandwidth B in radians:

$$R \approx e^{-B/2}$$

Show that this is very close to the result we got in Eq. 3.6, and try to explain where this exponential form comes from.

2. Fill in the steps between Eqs. 5.1 and 5.2 in the derivation of the exact resonance frequency of a reson filter. Then finish the derivation of Eq. 5.3.

3. (a) For what ranges of values of R and θ is the discrepancy greatest between the true peak frequency θ_0 and the pole angle θ?

(b) When θ is small, is θ_0 shifted to higher or lower frequencies?

(c) Derive an estimate for the discrepancy in terms of θ and $1 - R$.

(d) Try to synthesize an example in which the difference is audible.

4. (a) Show that the gain factor to use if we want to normalize by the magnitude response at the pole angle θ instead of the peak resonant frequency θ_0 is

$$A = (1 - R)(1 - 2R\cos(2\theta) + R^2)^{1/2}$$

(b) Show that this gain A approaches the gain A_0 in Eq. 5.5 as R approaches 1.

5. Show from Eq. 5.3 for a two-pole reson that for any $R > 0$ there is always a pole angle θ that achieves a specified true peak frequency ψ. Show on the other hand that for some combinations of values for R and θ there may be no peak frequencies other than at 0 or π.

6. Suggest some applications for which it would be useful to keep the bandwidth of a reson fixed when we change its center frequency. Then try to think of applications for which we would want to change the bandwidth when we change center frequency. What about situations in which we would want to change bandwidth but keep center frequency fixed?

7. As mentioned in Section 7 and the Notes, Smith and Angell propose reson_z mainly to keep the peak gain (magnitude response) close to constant as θ is changed and R is kept constant. They suggest putting the zeros at $\pm\sqrt{R}$ as an alternative to ± 1. Call the result *reson_R*, to distinguish it from reson_z.

(a) Write the transfer function $\mathcal{H}(z)$ of reson_R.

(b) Write the equation giving the output y at time t in terms of past and present inputs x and past outputs.

(c) Show that the magnitude response at the frequency corresponding to angle θ does not depend on the pole angle θ, but only on the radius R.

8. Here are some nitpicking details I point out in the article mentioned in the Notes. Don't work on this problem unless you have nothing else to do.

The actual peaks in the magnitude response of reson, reson_z, and reson_R do not occur precisely at the pole angle θ. This means that the peak gain of reson_R isn't actually constant when θ is changed and R is kept constant.

(a) Find the true peak frequency ψ of reson_z in terms of θ and R. Is there always a ψ given θ? A θ given ψ?

(b) Find the true peak gain of reson_z. The answer is actually independent of θ!

(c) The answer to (b) means that reson_z is preferable to reson_R if we want to keep the peak gain constant while changing θ. Why?

9. As mentioned in the Notes, the terms ''feedback filter'' and ''IIR filter'' are not perfectly synonymous. Give a concrete example of a feedback filter that is not an IIR filter. What about an IIR filter that is not a feedback filter?

10. The argument leading to Eq. 2.7, showing that a feedback filter is stable if and only if all its poles lie inside the unit circle, is heuristic and not rigorous. Criticize it. Fix it.

CHAPTER 6

Comb and String Filters

1 Comb filters

In this chapter we're going to explore some filters that are very useful for generating and transforming sound. By the end of the chapter we will have built an astonishly simple and efficient digital plucked-string instrument that is used all the time in computer music. In the process you'll get some practice using feedback filters, and also build up some intuition relating the behavior of comb filters to the standing waves on strings and in columns of air that we studied in Chapter 2.

In Chapter 4 we looked at inverse comb filters, but I haven't said yet where that name comes from. The inverse comb is described by the filter equation

$$y_t = x_t - R^L x_{t-L} \tag{1.1}$$

where x_t and y_t are the input and output signals, as usual. This filter uses only past inputs — no past outputs — so it's a feedforward filter. Suppose instead we consider the feedback filter that uses the past output y_{t-L} in place of $-x_{t-L}$:

$$y_t = x_t + R^L y_{t-L} \tag{1.2}$$

The reason we changed the sign of the delayed output signal will become clear shortly. To find the transfer function and frequency response, write Eq. 1.2 symbolically:

$$Y = X + R^L z^{-L} Y \tag{1.3}$$

Solving for the transfer function Y/X gives

$$\mathcal{H}(z) = \frac{1}{1 - R^L z^{-L}} \tag{1.4}$$

We'll call this feedback filter a *comb filter*.

Notice that the transfer function of this feedback filter is precisely the reciprocal of the transfer function of the feedforward filter called an inverse comb in Section 8 of Chapter 4. From this it follows that the magnitude frequency response of the comb filter is the reciprocal of that of the inverse comb. This explains the terminology. In fact, if we follow one filter with the other, the net result is to restore the original input signal; the two filters cancel each other out. Let's check this for the case of an inverse comb followed by a comb. As above, call the input to the inverse comb x and its output y. The signal y then becomes the input to the comb; call its output w. The equations for the two filters then become

$$y_t = x_t - R^L x_{t-L}$$
$$w_t = y_t + R^L w_{t-L}$$
(1.5)

Solve the second equation for y_t and substitute in the first, yielding

$$x_t - R^L x_{t-L} = w_t - R^L w_{t-L}$$
(1.6)

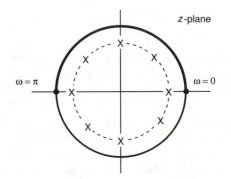

Fig. 1.1 The location of the poles of the comb filter in Eq. 1.2. The plot shown corresponds to a loop delay of $L = 8$ samples, and hence shows 8 poles.

Our goal is to show that the signals x and w are identical. Before we rush to conclude that this is implied by Eq. 1.6, there's a detail I've glossed over. We have to say something about how the filters get started, and whether the input x has been going on for all time. Let's make the simple assumption here that the signal x doesn't get started until $t = 0$ — in other words, that x_t is zero for $t < 0$. (I'll leave the fine point of what happens otherwise for Problem 1.) If this is so then Eq. 1.6 implies that $x_t = w_t$ for $t = 0, 1, \ldots, L-1$, because the delayed terms x_{t-L} and w_{t-L} are zero in that range. This implies, by the same reasoning, that $x_t = w_t$ for $t = L, L+1, \ldots, 2L-1$. We can continue the argument to infinity, block of L by block of L. In fancier terminology, this is a proof by induction on blocks of signal of length L.

Figure 1.1 shows the poles of the comb filter described by Eq. 1.2. They're exactly where the zeros of the inverse comb are (see Fig. 8.1 in Chapter 4) — at the zeros of the denominator $1 - R^L z^{-L}$, equally spaced around the circle of radius R.

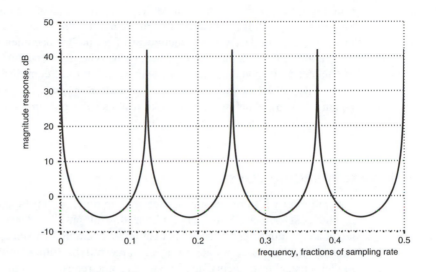

Fig. 1.2 Frequency response of the comb filter described in the previous figure. The case shown is for $R = 0.999$.

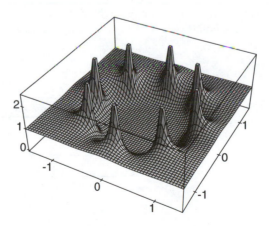

Fig. 1.3 Magnitude of the transfer function of the comb of Figs. 1.1 and 1.2, shown as a contour plot above the z-plane.

You can view the canceling of the comb and inverse comb filters as simply the poles of the comb transfer function canceling the zeros of the inverse comb transfer function. Figure 1.2 shows the magnitude frequency response of the comb for the case of a loop delay $L = 8$. It is of course just the reciprocal of the magnitude response of the inverse comb. In decibels, the reciprocal of a number becomes its negative, so one magnitude plot can be obtained by turning the other upside down, as hinted at the end of Chapter 4. It should be clear now why we call these ''comb'' filters.

Finally, a bird's-eye view of the magnitude response as a contour plot above the z-plane is shown in Fig. 1.3. It looks just like the pole pairs of three resons, plus another pair of poles at ± 1.

2   Analogy to standing waves

A comb filter works by adding, at each sample time, a delayed and attenuated version of the past output. You can think of this in physical terms: The delayed and attenuated output signal can be thought of as a returned *traveling wave*. This kind of analogy has been used in interesting ways recently by Julius Smith and Perry Cook to model musical instruments, and I've given some references to their work in the Notes at the end of this chapter. I want to introduce just the flavor of the idea here, and I'll return to this theme later when we discuss reverberation in Chapter 14.

Figure 2.1 shows the signal flowgraph for a comb filter, with some suggestion of a traveling-wave interpretation. An output wave travels around the feedback loop and returns after a delay of L samples; the return wave is added to the input, but only after it is attenuated by the factor R^L, which we'll assume is less than one in magnitude. You can think of the parameter R as the signal attenuation per sample. For example, the wave may be traveling through air, which absorbs a fraction of the wave's energy every T_s seconds, where T_s is the sampling interval. The delay L is the *round-trip* time in samples.

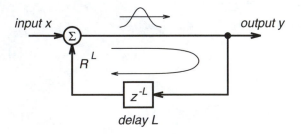

Fig. 2.1 The signal flowgraph of a comb filter with a hint of its traveling-wave interpretation.

In vibrating columns of air, waves are reflected at boundaries in two ways. Reflection at a closed end of a tube inverts the sign of a wave; reflection at an open end doesn't. What matters in Fig. 2.1 is the net effect of a round-trip, which we want to be no change in sign, being that we've chosen the sign of the feedback term to be positive. Therefore the appropriate analogy to the comb filter is a tube of length $L/2$ (so the round-trip time is L), either closed at both ends or open at both ends. As for strings, we can't really imagine a vibrating string that is not tied down at its ends, so the analogy must be a string fixed at both ends, also of length $L/2$.

A string tied down at both ends (or a tube open or closed at both ends) has the natural resonant frequencies $k2\pi/L$ radians per sec, as shown in Eq. 5.15 in Chapter 2, where k is any integer. (That equation actually has frequencies $k\pi/L$; the factor of two is explained by the fact that here the string is of length $L/2$ instead of L.) These are precisely the angles where the poles of the comb filter occur in the z-plane. This checks our interpretation and gives us an intuitive way to understand the resonances as standing waves. The resonant frequencies are the ones that fit in the feedback loop an integer number of times, just as the standing waves are the waves that fit perfectly on the string or in the tube.

What happens if the sign of the wave is inverted in the course of a round-trip of the feedback loop? This corresponds to replacing the plus sign in Eq. 1.2 by a minus:

$$y_t = x_t - R^L y_{t-L} \tag{2.1}$$

Physically, this corresponds to the vibration of air in a tube that is closed at one end and open at the other. Recall from our work with tubes that the fundamental resonant frequency is now half of what it was with both ends closed or open, and that only odd harmonics are possible (see Eq. 7.12 in Chapter 2). Algebraically these frequencies are $k2\pi/(2L) = k\pi/L$, where k is an odd integer. (Again, there is a factor of two because now the round-trip length is L instead of $2L$.) The physical picture has sinusoids with maxima or minima at one end and zeros at the other (Fig. 7.1 in Chapter 2).

Let's check the comb filter with a sign inversion against the tube closed at one end and open at the other. The transfer function corresponding to Eq. 2.1 is

$$\mathcal{H}(z) = \frac{1}{1 + R^L z^{-L}} \tag{2.2}$$

and the poles are at roots of the equation

$$z^L = -1 \tag{2.3}$$

Since

$$-1 = e^{j\pi} \tag{2.4}$$

the roots are all shifted by an angle π/L with respect to the case of a comb filter without sign inversion, as shown in Fig. 2.2. These pole angles are in fact odd harmonics of the fundamental frequency π/L. I hope by this point you can anticipate the frequency response, shown in Fig. 2.3, from the pole pattern. Each pole near the unit circle causes a resonance peak — and the resonant frequencies of the comb are exactly the same as those of the analogous resonant tube.

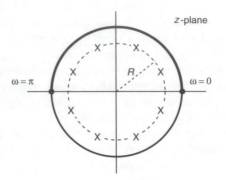

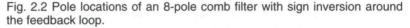

Fig. 2.2 Pole locations of an 8-pole comb filter with sign inversion around the feedback loop.

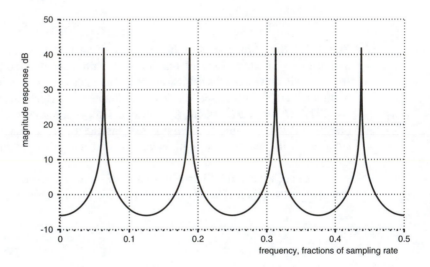

Fig. 2.3 Frequency response of the 8-pole comb filter with sign inversion. The case shown is for $R = 0.999$.

3 Plucked-string filters

Comb filters are versatile building blocks for making sounds of all sorts. Variations of them can be used to simulate reverberation, to transform the character of any input sound, or to construct a very striking and efficient instrument that sounds like a plucked string.

A comb filter gets you a lot for a little — a simple delay line holding 50 samples results in a filter with 25 pole-pairs, and therefore 25 resonant frequencies. The filter's frequency response is complicated, but its implementation entails only one multiplication and one addition per sample. In a sense the comb's success stems from the fact that it models a basic physical phenomenon: the return of an echo.

Next, I want to show how the physical interpretation of a comb filter can be exploited to derive the plucked-string filter. Suppose we apply a unit impulse to the comb filter in Eq. 1.2. What is the resulting impulse response? Well, the unit impulse returns L samples later, and is multiplied by the coefficient R^L. There is no other input to the delay line, so nothing else happens between time 0 and time L. The pulse of height R^L then enters the delay line and returns L samples later, and so forth. The impulse response is therefore

$$h_t = \begin{cases} R^t & \text{if } t = 0 \bmod L \\ 0 & \text{elsewhere} \end{cases} \tag{3.1}$$

That is, $h_t = R^t$ for $t = 0, L, 2L, \ldots$, and zero otherwise. This is a periodic sequence of pulses at the fundamental frequency f_s/L Hz, an integer fraction of the sampling rate — except that it decays at a rate determined by the parameter R. The closer R is to one, the slower it decays (and the closer the poles are to the unit circle). If you listen to the impulse response of a comb filter, that's exactly what you hear: a buzzy sound with pitch f_s/L that dies away. Not very exciting.

Remember that a string tied down at both ends supports standing waves because traveling waves are reflected from both those ends. The behavior of a comb filter is very similar, as noted in the previous section. We might therefore expect the impulse response of a comb filter to sound like a string that is suddenly set in motion — but it doesn't. Why not? Because the sound of a string changes in a certain way over the course of a note. This is an example of a recurrent theme in computer music and in psychoacoustics in general: sounds are not interesting unless they change their frequency content with time. Perfectly periodic waveforms are boring.

But the behavior of the comb filter does reflect the fact that a plucked string does not go on vibrating forever. Its energy is gradually dissipated to the rest of the world. Some energy is radiated in the form of sound to the air, and some is transmitted to the supports at the ends of the string, where it contributes to the radiation of sound by other parts of the instrument, like the bridge and sounding board. This decay in energy is captured by the poles of the comb filter being inside the unit circle at radius R, causing the waveform amplitude to decay by the factor R every sample (R^L in L samples).

We're still missing something critical: the insight that the different frequency components produced by a vibrating string decay *at different rates* [Karplus and Strong, 1983]. The high frequencies die away much faster than the low frequencies. This is illustrated in Fig. 3.1, which shows the spectrogram of a real plucked string, an acoustic guitar note. Notice how the high-frequency components present at the attack die away faster than the fundamental components.

Karplus and Strong suggest a very clever modification of the comb filter to take this effect into account. The idea is to insert a lowpass filter in the feedback loop so that every time the past output signal returns, its high-frequency components are diminished relative to its low-frequency components. This works like a charm. In fact it works so well it seems almost like magic.

What's more, the following very simple lowpass filter works well:

$$y_t = \tfrac{1}{2}[x_t + x_{t-1}] \tag{3.2}$$

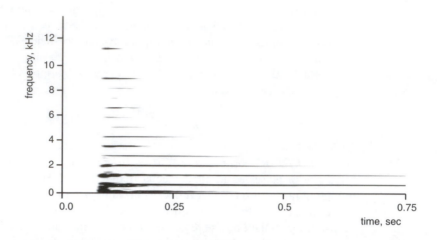

Fig. 3.1 Spectrogram of a real acoustic guitar note, the F# in the octave above middle C's octave, which corresponds to a pitch of 740 Hz. The abscissa is time in sec, and the ordinate is frequency in kHz. The horizontal bars show the harmonics of the fundamental frequency.

Except for the factor of two, we looked at the same filter in Section 7 of Chapter 4; it has the transfer function

$$\mathcal{H}(z) = \tfrac{1}{2}[1 + z^{-1}] \tag{3.3}$$

with a zero at the point in the z-plane that corresponds to the Nyquist frequency, $z = -1$. Its complex frequency response can be put in the form

$$H(\omega) = \cos(\omega/2)\, e^{j\omega(t - 1/2)} \tag{3.4}$$

The magnitude response $|H(\omega)| = |\cos(\omega/2)|$ starts at unity for zero frequency and slopes down to zero at the Nyquist frequency, as shown in Fig. 3.2. This is a modest lowpass filter, but it does the job nicely. The intuition is that the filter operates on the signal every time it executes a round-trip around the feedback loop of the comb. After m round-trips, the signal has been processed by the lowpass filter m times, so its frequency content has been multiplied by $|H(\omega)|^m$. It is therefore a good thing that the filter has a gently sloping frequency response; otherwise, the high-frequency components of the signal would get wiped out too fast.

Figure 3.3 shows the spectrogram of a note produced by the plucked-string filter just described. The abscissa shows time in sec, and the ordinate shows the frequency content for a time segment around that time. The faster decay of the higher frequencies is just what we planned.

We now come to an interesting point about phase response. In many cases, the phase response of a filter is not critical, but when the filter is in a feedback loop, as it is now, its effect on phase *can* be critical. Equation 3.4 shows that the lowpass filter has linear phase, so its delay is the same for all frequencies. In fact, the complex

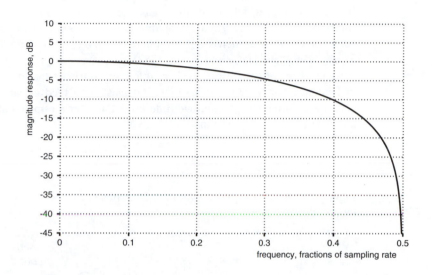

Fig. 3.2 The magnitude response of the simple lowpass feedforward filter used in the plucked-string filter.

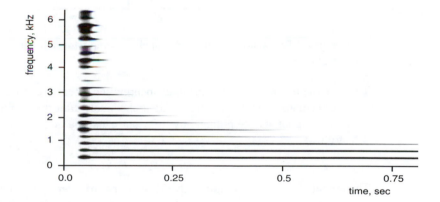

Fig. 3.3 Spectrogram of a note produced by the plucked-string filter. The abscissa shows time in sec, and the ordinate shows frequency in kHz. The parameters are $R = 0.9995$, $L = 75$, and $f_s = 22,050$. One hundred random numbers were used as initial input.

exponential factor is precisely equivalent to a delay of one-half sample. The loop delay is therefore $L + \frac{1}{2}$ samples, *not* L samples, and the fundamental frequency generated is $f_s/(L + \frac{1}{2})$. This is not a trivial matter; when $L = 50$, for example, the difference in frequency caused by the lowpass filter is about 1 percent — easily discernible by ear.

The plucked-string filter we have now is so nice to listen to, and so efficient, that it is one of the most commonly used computer instruments for real-time applications. Because it is so widely used, there has been a fair amount of work in tuning it up (literally and figuratively), and extending it to other kinds of sounds. We will discuss some of these ideas here, and for more information you should see [Karplus and Strong, 1983] and [Jaffe and Smith, 1983].

The filter we've constructed is a little more intricate than the simple feedforward or feedback filters we've seen so far: It consists of a feedforward loop within a feedback loop. In the next section we'll describe the filter's implementation and then take a look at its frequency response.

4 Implementing plucked-string filters

Figure 4.1 shows a signal flowgraph of the plucked-string filter, with its feedforward loop within its feedback loop.

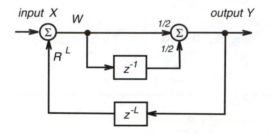

Fig. 4.1 Signal flowgraph for the plucked-string filter. Note the intermediate signal w.

To write the update equations for implementing the filter, it's convenient to introduce the intermediate signal w, which appears immediately after the closing of the feedback loop. The signal w is determined by the input x and the delayed and weighted output y, as follows:

$$w_t = x_t + R^L y_{t-L} \tag{4.1}$$

The output at time t is determined by the feedforward filter with input w, so

$$y_t = \tfrac{1}{2} w_t + \tfrac{1}{2} w_{t-1} \tag{4.2}$$

This is a little different from the situations we've seen up to now. The determination of the next value of the output is determined by two equations instead of one. But this presents no new difficulty. At each value of the sample number t, we first find w_t from x_t and y_{t-L}, using Eq. 4.1; then we find the output value y_t from w_t and w_{t-1}, using Eq. 4.2. Of course, just as in the simple feedforward or feedback filter, we need to save signal values for future use. In this case we need to save the past output values back to y_{t-L}, as well as the value of w at the previous sample, w_{t-1}.

You should be a little worried at this point about the possible side effects of what we did. We inserted a lowpass filter in the feedback loop of a comb to attenuate the high frequencies as they circulate around the loop. The magnitude response of the lowpass filter does have the desired effect, as we've seen from the spectrogram in Fig. 3.3. The phase response of the lowpass filter introduces an additional half-sample delay, and we argued that this makes the loop delay $L + \frac{1}{2}$ samples instead of L. But how do we know where the resonances of the altered filter really are? Are they at multiples of the fundamental frequency $f_s/(L + \frac{1}{2})$? The resonances of a filter with feedback are determined by its poles, and in the case of the simple comb, the poles are at the Lth roots of unity — equally spaced in frequency. But now the algebraic determination of the pole locations is very difficult (I don't know if it's even possible), and we are forced to look directly at the frequency response.

5   Resonances of the plucked-string filter

To look at the frequency response of the plucked-string filter we need to derive its transfer function. This is not very hard, using the same symbolic approach we used for simpler filters. Recall that a delay of one sample is represented by the operator z^{-1}; a delay of L samples by z^{-L}. In terms of these operators, Eqs. 4.1 and 4.2 become

$$W = X + R^L z^{-L} Y \tag{5.1}$$

and

$$Y = \frac{1}{2}[1 + z^{-1}]W \tag{5.2}$$

It is now a matter of a little algebra to solve for the ratio Y/X, the transfer function of the filter from input X to output Y. First substitute the expression for W in Eq. 5.1 into Eq. 5.2, getting an equation involving only Y and X. Then solve for Y in terms of X, yielding

$$\mathcal{H}(z) = Y/X = \frac{\frac{1}{2}[1 + z^{-1}]}{1 - R^L z^{-L} \frac{1}{2}[1 + z^{-1}]} \tag{5.3}$$

We're most interested in the magnitude response corresponding to this transfer function. This is not really hard to compute, but I want to take a little time to explain some details of the program I wrote to do it. It will be a good review of the previous two chapters. First, I multiplied the numerator and denominator of Eq. 5.3 by $2z^{L+1}$, to get the transfer function in the less confusing and more conventional form of a ratio of polynomials:

$$\mathcal{H}(z) = \frac{z^{L+1} + z^L}{2z^{L+1} - R^L z - R^L} \tag{5.4}$$

We want to evaluate this for z on the unit circle, so I then replaced z and its powers using Euler's formula:

$$\begin{aligned} z &= \cos\omega + j\sin\omega \\ z^L &= \cos(L\omega) + j\sin(L\omega) \\ z^{L+1} &= \cos((L+1)\omega) + j\sin((L+1)\omega) \end{aligned} \tag{5.5}$$

It's then easy to write out explicitly the real and imaginary parts of the numerator and denominator:

$$\begin{aligned}
\mathscr{Real}\ \{\text{numerator}\} &= \cos((L+1)\omega)\ +\ \cos(L\omega) \\
\mathscr{Imag}\ \{\text{numerator}\} &= \sin((L+1)\omega)\ +\ \sin(L\omega) \\
\mathscr{Real}\ \{\text{denominator}\} &= 2\cos((L+1)\omega)\ -\ R^L\cos\omega\ -\ R^L \\
\mathscr{Imag}\ \{\text{denominator}\} &= 2\sin((L+1)\omega)\ -\ R^L\sin\omega
\end{aligned} \tag{5.6}$$

where $\mathscr{Real}$ and $\mathscr{Imag}$ denote the real and imaginary parts, respectively. I assigned temporary variables for these four components, the real and imaginary parts of the numerator and denominator. The magnitude response is the magnitude of this as a complex function, or

$$|H(\omega)|\ =\ \left[\frac{[\mathscr{Real}\ \{\text{numerator}\}]^2\ +\ [\mathscr{Imag}\ \{\text{numerator}\}]^2}{[\mathscr{Real}\ \{\text{denominator}\}]^2\ +\ [\mathscr{Imag}\ \{\text{denominator}\}]^2}\right]^{\frac{1}{2}} \tag{5.7}$$

I then just evaluated this for ω on a grid in the range from 0 to π radians per sample.

Figure 5.1 shows the result when $L = 32$ and the coefficient $R = 0.999$. Since the round-trip delay of the feedback loop is 32.5 samples, we expect the resonances to occur at integer multiples of $f_s/32.5$, and these frequencies are marked by triangles on the graph.

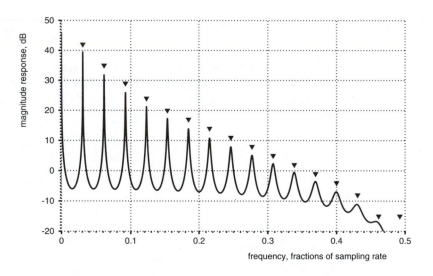

Fig. 5.1 Magnitude response of a plucked-string filter, for the case of a loop length $L = 32$ and pole radius $R = 0.999$. The expected resonant frequencies, integer multiples of $f_s/32.5$, are indicated by triangles.

The first interesting thing to notice in Fig. 5.1 is that the resonance peaks increase in width as frequency increases. This is exactly what we should expect, since the lowpass filter inserted in the loop causes higher frequencies to decay faster. Wider resonance peaks correspond to poles farther from the unit circle, and to signal components that decay faster.

Second, the peaks in the magnitude response line up very closely with the predicted integer multiples of $f_s/32.5$. The peaks are not precisely at the expected points, and they also are not at exact integer multiples of the frequency of the first peak. This deviation of the overtone series from a simple harmonic progression is smaller when the harmonic numbers are lower, and also when the loop delay L is larger (corresponding to a lower fundamental frequency). But the deviations, especially at the lower harmonics, are very tiny. Tinier than we usually need to worry about. For example, in our example with a loop delay of 32 samples, the tenth harmonic is off by only 0.027 percent. The deviations for lower harmonics and lower-pitched filters are even smaller.

A third noticeable difference in the magnitude response of the plucked-string filter, compared to a comb filter, is its general lowpass shape. The peaks decrease in amplitude with increasing frequency, whereas the peaks of the comb filter are all of equal height. This is not surprising, since we inserted a lowpass filter in the path between input and output.

The plucked-string filter is so useful for musical purposes that we will want to be able to tune its pitch very finely. That leads us to the first-order allpass filter, a useful and interesting filter in its own right.

6 The first-order allpass filter

At this point we have only crude control over the pitch of a plucked-string filter. We can choose the integer loop length L, yielding a fundamental pitch $f_s/(L + \frac{1}{2})$, but that integer L is all we have to work with. To see just how crude this control of pitch is, let's see what happens when $L = 10$. This is a perfectly reasonable example, by the way; if the sampling rate is 22,050 Hz, a loop length of 10 corresponds to a pitch of $22{,}050/10.5 = 2100$ Hz, which is very close to the C three octaves above middle C. Now suppose we decrease L by one. This increases the pitch to $22{,}050/9.5 = 2321.05$ Hz, which is almost up to the following D, a jump of almost a full step in the scale. We appear to be in real trouble if we want to produce the C# between the two. Smooth glissandos seem to be out of the question. Getting better control over the pitch of the plucked-string filter presents an interesting problem, which we'll now address.

Intuitively, the fundamental resonant frequency of the plucked-string filter is determined by the total delay around the feedback loop. If the total delay is D samples, or DT_s sec, the first resonant frequency is $1/(DT_s) = f_s/D$ Hz. We haven't said anything about D being an integer number of samples. In fact, in the plucked-string filter we have so far, D is the sum of the integer buffer length, L, plus one-half sample due to the lowpass filter, so D is *not* an integer. What we would like is a way to introduce additional delays of *fractions* of a sample period in the feedback loop. That would enable us to fine-tune the delay D and hence the pitch.

In fact, what we'd like is a filter that introduces, or comes close to introducing, an arbitrary fractional delay, but has no effect on the magnitude of the frequency

response around the feedback loop. We already have a loop with the lowpass characteristic we want for the plucked-string sound, and we don't want to tamper with a
good thing. The idea is to try to construct a filter that has no effect on the magnitude
of phasors, no matter what their frequency. Suppose we start with a pole at $z = \rho$,
where ρ is some real number. Maybe we can add a zero to the filter transfer function
so that the effect of the pole on the magnitude response will be canceled. Where
should we put the zero? One answer is: the same place — that will cancel the effect
of the pole perfectly. But, of course, that accomplishes nothing; it gets us back to a
unity transfer function, and has no effect on the phase response.

Putting the zero at $-\rho$ doesn't do the trick. If ρ is positive, for example, the pole
will have a lowpass effect, and a zero at $-\rho$ will have the same effect. The result will
be to exaggerate rather than cancel the effect of the pole.

There aren't many other places to try. How about putting the zero at $z = 1/\rho$?
That does put the zero closer to the lower than the higher frequency points on the unit
circle, so its effect will be highpass — opposite that of the pole. This sounds promis-
ing. Let's look at the magnitude response, using Fig. 6.1. The vector from the pole to
an arbitrary point on the unit circle is labeled with length B, and the corresponding
vector from the zero is labeled with length C. The point on the unit circle is at frequency θ radians per sample.

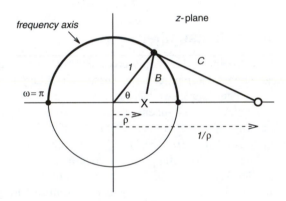

Fig. 6.1 Pole-zero diagram and some geometry for the first-order allpass
section.

Recall that the magnitude response at frequency θ due to a zero is the length of the
vector from the zero to the point on the unit circle at angle θ (Section 6 of Chapter 4).
Similarly, the magnitude response due to the pole is the reciprocal of the length of the
vector from the pole to the point on the unit circle. Therefore the magnitude response
of the filter with both the zero and pole is the ratio of these lengths, C/B. Let's try to
put this in terms of θ and the constant ρ.

We can express B in terms of ρ and θ using the law of cosines:

$$B^2 = 1 + \rho^2 - 2\rho\cos\theta \tag{6.1}$$

Similarly, we can write C in terms of ρ and θ using the same law:

$$C^2 = 1 + 1/\rho^2 - 2(1/\rho)\cos\theta \qquad (6.2)$$

Multiplying Eq. 6.2 by ρ^2 yields the right-hand side of Eq. 6.1, so

$$\rho^2 C^2 = B^2 \qquad (6.3)$$

or, forming the square of the magnitude response C/B:

$$C^2/B^2 = 1/\rho^2 \qquad (6.4)$$

This is even better than we could have hoped for: The magnitude of the frequency response is perfectly independent of frequency! This sounds almost magical, but it is correct: if you place a zero at the reciprocal of the real pole position, the filter has a magnitude response that is absolutely constant with respect to frequency. All frequencies are passed with equal weight. We call such filters *allpass* filters.

Before we go on, remember that it is perfectly acceptable to have a zero outside the unit circle. A pole outside the unit circle causes instability, as noted in Section 2 of Chapter 5. But zeros are tamer creatures, and we can put them anywhere in the z-plane. This makes the allpass construction feasible.

We want to look at the phase response of our single-pole, single-zero allpass filter, but first let's construct the transfer function corresponding to the pole and zero in Fig. 6.1:

$$\mathcal{H}(z) = K\,\frac{z + 1/a}{z + a} \qquad (6.5)$$

where K is any constant, and we've used the parameter a to avoid minus signs; the pole is at the point $z = -a$. We have the freedom to choose the constant factor K any way we want; it is convenient to choose it to force the transfer function to have the value one at zero frequency, the point $z = 1$. Setting $\mathcal{H}(1) = 1$ gives us $K = a$, and hence the transfer function is

$$\mathcal{H}(z) = \frac{z^{-1} + a}{1 + az^{-1}} \qquad (6.6)$$

As usual, we've written the transfer function in terms of z^{-1}, the delay operator.

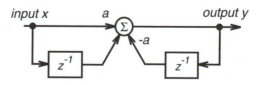

Fig. 6.2 Signal flowgraph for a first-order allpass filter.

The allpass filter we've just derived is a combination feedforward and feedback filter. If we choose to implement the feedforward part before the feedback part, we get the signal flowgraph shown in Fig. 6.2, corresponding to the filter equation

$$y_t = ax_t + x_{t-1} - ay_{t-1} \tag{6.7}$$

See Problem 5 for a more efficient implementation.

7 Allpass phase response

We finally get to the phase response of the allpass filter, the reason we started looking at it in the first place. When we get its phase response ϕ, it will tell us how much a phasor of frequency ω applied to the filter will be delayed. To be specific, suppose we apply the phasor $e^{j\omega t}$ as input. The output signal will be the phasor of unit magnitude with its phase shifted by $\phi(\omega)$:

$$e^{j\omega t}e^{j\phi(\omega)} = e^{j\omega(t + \phi(\omega)/\omega)} \tag{7.1}$$

The right-hand side of Eq. 7.1 shows that the phasor is shifted by $\phi(\omega)/\omega$ samples. The phase response $\phi(\omega)$ is usually negative, so $-\phi(\omega)/\omega$ represents a delay, called the *phase delay*.[†] In general, this phase delay is a function of the frequency ω. All this checks with the discussion in Section 7 of Chapter 4, where we pointed out that *exactly* linear phase in a feedforward filter results in a constant delay. What we're looking for in the allpass filter is a phase response that is at least approximately linear. Remembering that, let's find $\phi(\omega)$ for the allpass filter.

The way to start calculating either the magnitude or the phase response is to replace z by $e^{j\omega}$ in the transfer function, Eq. 6.6, to get the frequency response

$$H(\omega) = \frac{e^{-j\omega} + a}{1 + ae^{-j\omega}} \tag{7.2}$$

We could now find the phase response $\phi(\omega)$ by finding the real and imaginary parts of the numerator and denominator, and using the arctangent function as follows

$$\phi(\omega) = \arctan\left[\frac{\mathcal{Imag}\,\{\text{numerator}\}}{\mathcal{Real}\,\{\text{numerator}\}}\right] - \arctan\left[\frac{\mathcal{Imag}\,\{\text{denominator}\}}{\mathcal{Real}\,\{\text{denominator}\}}\right] \tag{7.3}$$

in analogy to the magnitude calculation we did in Section 5. Instead, I'm going to be a little tricky, in order to get the result in a particularly convenient form.

The idea is to try to introduce some symmetry in Eq. 7.2 by multiplying the numerator and denominator by $e^{j\omega/2}$:

$$H(\omega) = \frac{e^{-j\omega/2} + ae^{j\omega/2}}{e^{j\omega/2} + ae^{-j\omega/2}} \tag{7.4}$$

A good thing has now happened: We've succeeded in making the denominator very similar to the numerator. In fact, the only difference between them is that one is the complex conjugate of the other. If we set the numerator to $re^{j\psi(\omega)}$, the denominator is $re^{-j\psi(\omega)}$, and the ratio can be written

$$H(\omega) = \frac{re^{j\psi(\omega)}}{re^{-j\psi(\omega)}} = e^{2j\psi(\omega)} \tag{7.5}$$

[†] The term *phase delay* is used to distinguish this from *group delay*. See the Notes at the end of this chapter.

This shows that the phase response $\phi(\omega)$ is simply $2\psi(\omega)$, twice the phase angle of the numerator in Eq. 7.4. (As a side effect, it also confirms that the magnitude of the transfer function is unity.) The numerator can be written

$$(a + 1)\cos(\omega/2) + j(a - 1)\sin(\omega/2) \tag{7.6}$$

so the phase response of the allpass, finally, is

$$\phi(\omega) = -2\arctan\left[\frac{1 - a}{1 + a}\tan(\omega/2)\right] \tag{7.7}$$

This form for the phase is compact and pretty, but it's also particularly illuminating if we focus our attention on low frequencies. When x is small, $\tan x \approx x$, and this gives us the following low-frequency approximation

$$\phi(\omega) \approx -\frac{1 - a}{1 + a}\omega \approx -\delta\omega \tag{7.8}$$

where we've defined

$$\delta = \frac{1 - a}{1 + a} \tag{7.9}$$

The variable δ is an approximation to the phase delay $-\phi(\omega)/\omega$. From our discussion at the beginning of this section, the phase delay of the allpass filter is approximately equal to δ for low frequencies. We can also solve for a in terms of δ:

$$a = \frac{1 - \delta}{1 + \delta} \tag{7.10}$$

which is a handy formula if we specify the phase delay.

In practice a must always be less than one (why?), so δ is always positive. Furthermore, there is not much point in trying to approximate delays greater than one sample with the allpass, because we can always take care of the integer part by absorbing it into the buffer used to implement the loop delay, the integer L. We can therefore restrict δ to the range between 0 and 1, which is equivalent to restricting a to the same range.

Figure 7.1 shows plots of the phase response of the allpass filter for the ten values of a corresponding to $\delta = 0.1, 0.2, \ldots, 1.0$ samples. As predicted, the phase looks linear at low frequencies, with slope approximately equal to $-\delta$. The phase delay $-\phi(\omega)/\omega$ gives us a better idea of the quality of the approximation, and is plotted in Fig. 7.2 for the same range of δ. We see that the allpass delivers close to the desired delay at low frequencies. The errors are quite small for frequencies below $0.05f_s$. At the frequency $0.05f_s$, for example, which is 1102.5 Hz at a sampling rate of 22,050, the error is only 0.0031 samples for $\delta = 0.5$ samples. At the higher frequency of $0.2f_s$, the error is up to 0.0546 samples at the same δ.

Notice also from Fig. 7.2 that the allpass filter's approximation to constant delay is better for values of delay near 0 or 1 sample than it is near 0.5 samples. Think of it this way: a delay of a fraction of a sampling interval actually interpolates the signal between sample values. Interpolating midway is most difficult, because that point is farthest from known sample values. Still, the one-zero, one-pole allpass filter does a reasonably good job at all delays for low frequencies.

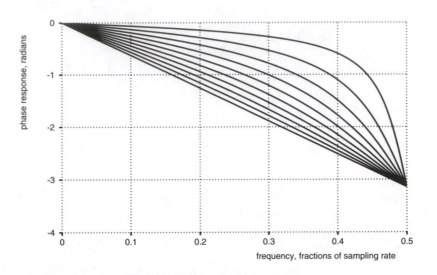

Fig. 7.1 Phase response for the first-order allpass filter; from the top, the prescribed delays δ are 0.1, 0.2, ... ,1.0 samples.

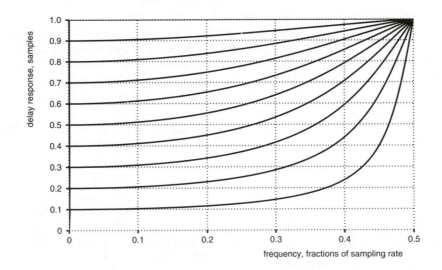

Fig. 7.2 Delay response for the first-order allpass filter; from the bottom, the prescribed delays δ are 0.1, 0.2, ... ,1.0 samples.

All our work on the allpass filter has paid off. We've shown that it provides an effective and efficient way to tune the delay in a feedback loop. Now let's put the plucked-string instrument together.

8 Tuning plucked-string filters

Figure 8.1 shows the finished plucked-string filter. We have two main points left to discuss: the final details of tuning and the selection of the input signal.

The tuning of all the harmonics simultaneously with the allpass filter is not possible; from the plot of phase delay we see that the upper harmonics will have greater relative delay than the fundamental. That is, the upper partials will be flat. Jaffe and Smith [1983] suggest that this is not such a bad thing perceptually, and recommend tuning the filter so that the fundamental frequency is exactly correct. Let's run through an example to see how they do this.

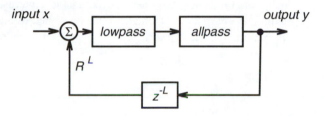

Fig. 8.1 Tunable plucked-string filter.

Suppose we want the lowest resonance of the plucked-string filter to occur at precisely 1000 Hz, using a sampling rate of 22,050 Hz. This corresponds to a loop delay of $22{,}050/1000 = 22.05$ samples. Remember that the loop delay due to the buffer and lowpass filter is $L + \frac{1}{2}$ samples, so we should choose $L = 21$ to keep the δ of our allpass filter in the range between zero and one sample. We then wind up with a desired phase delay of $\delta = 22.05 - 21.5 = 0.55$ samples.

At this point we could use the approximate formula for the allpass filter parameter a in terms of specified phase delay, Eq. 7.10. But it would be best if we could get the frequency of the fundamental resonance exactly right. To do this, we stipulate that the negative of the exact phase response at the frequency ω_0 (from Eq. 7.7), divided by the frequency ω_0, be equal to the desired phase delay δ:

$$\delta = \frac{2}{\omega_0} \arctan\left[\frac{1-a}{1+a}\tan(\omega_0/2)\right] \tag{8.1}$$

By a stroke of luck, we can solve this exactly for the allpass filter parameter a in terms of δ [Jaffe and Smith, 1983]:

$$a = \frac{\sin((1-\delta)\omega_0/2)}{\sin((1+\delta)\omega_0/2)} \tag{8.2}$$

Notice that for small ω_0, this reduces to the approximate formula $(1 - \delta)/(1 + \delta)$, as we would expect. Our example, with $\omega_0 = 2\pi \cdot 1000/22{,}050$ radians and $\delta = 0.55$ samples, results in $a = 0.292495$. The approximate formula for a, Eq. 7.10, yields a phase delay of 0.552607 samples instead of the target 0.55 samples, about 0.5 percent high. Of course, the relative error in terms of the total loop delay of 22.05 samples is much smaller.

There are quite a few twists on the basic plucked-string filter idea, many mentioned in [Karplus and Strong, 1983] and [Jaffe and Smith, 1983]. I'll mention some of them in the Problems. But we've skipped a basic one, which I'll mention here: The initial excitation of the filter should be chosen to provide lots of high frequencies. This lends verisimilitude to the resulting note, ostensibly because the initial vibration of a real string has a healthy dose of high-frequency energy. The usual way to accomplish this is to start with an initial burst of random numbers, which I did to produce Fig. 3.3.

The output of the plucked-string filter is almost, but not quite, periodic. In some sense its musical quality depends both on its being close to periodic (so that it has a pitch), and on its not being periodic (so that it's interesting). In the next chapter we're going to develop the mathematical tools we need to understand perfectly periodic signals, paving the way for dealing with completely general signals.

Notes

Julius Smith has pioneered work on applying waveguide analogies to computer music. The following is a useful general article, with lots of references to related work:

> J. O. Smith, "Physical Modeling using Digital Waveguides, *Computer Music Journal*, vol. 16, no. 4, pp. 74–91, Winter 1992.

Perry Cook has developed some striking applications based on these ideas. See, for example:

> P. R. Cook, "Tbone: An Interactive WaveGuide Brass Instrument Synthesis Workbench for the NeXT Machine," *Proc. International Computer Music Conf*, San Francisco, International Computer Music Association, pp. 297–300, 1991.

> P. R. Cook, "SPASM, a Real-Time Vocal Tract Physical Model Controller; and Singer, the Companion Software Synthesis System," *Computer Music Journal*, vol. 17, no. 1, pp. 30–44, Spring 1993.

The following back-to-back articles are a rich source of interesting ideas for extending and improving the basic plucked-string filter.

> [Karplus and Strong, 1983] K. Karplus and A. Strong, "Digital Synthesis of Plucked-String and Drum Timbres," *Computer Music Journal*, vol. 7, no. 2, pp. 43–55, Summer 1983.

[Jaffe and Smith, 1983] D. A. Jaffe and J. O. Smith, "Extensions of the Karplus-Strong Plucked-String Algorithm," *Computer Music Journal*, vol. 7, no. 2, pp. 56–69, Summer 1983.

Paul Lansky's pieces "Night Traffic" and "Sound of Two Hands" are intriguing examples of using comb filters in computer music. His "Now and Then" uses plucked-string filters. These and other pieces are on his CD *HomeBrew*, Compact Disc BCD 9035, Bridge Records, 1992. Lansky uses digital signal processing and algorithmic techniques in much of his music. He comments in the liner notes to this disc that these pieces ". . . are attempts to view the mundane, everyday noises of daily life through a personal musical filter."

Add nonlinear distortion and feedback to the plucked-string filter and you get a versatile digital version of a rock guitar. Charles Sullivan shows how in the following paper, which makes ingenious use of many of the ideas we've studied up to now:

C. R. Sullivan, "Extending the Karplus-Strong Algorithm to Synthesize Electric Guitar Timbres with Distortion and Feedback," *Computer Music Journal*, vol. 14, no. 3, pp. 26–37, Fall 1990.

The family of allpass filters mentioned in Problem 9 is derived in

[Fettweis, 1972] A. Fettweis, "A Simple Design of Maximally Flat Delay Digital Filters," *IEEE Trans. on Audio and Electroacoustics*, vol. AU-20, pp. 112–114, June 1972.

That paper actually provides a simple derivation of earlier results of J.-P. Thiran; for example

J.-P. Thiran, "Recursive Digital Filters with Maximally Flat Group Delay," *IEEE Trans. on Circuit Theory*, vol. CT-18, pp. 659–664, Nov. 1971.

The distinction between phase and group delay is discussed in

A. Papoulis, *The Fourier Integral and its Applications*, McGraw-Hill, New York, N.Y., 1962.

Problems

1. An inverse comb filter is followed by a comb filter with the same parameter, as in Eq. 1.5. Construct an input signal x for which the output w of the comb filter is different from x. Hint: From the discussion x must be nonzero for arbitrarily negative t.

2. Here's a project for video gamesters. Write an interactive flight simulator whose landscape is the magnitude response of a feedback filter over the z-plane. Don't try to go over a pole!

3. [Karplus and Strong, 1983], [Jaffe and Smith, 1983] Is the plucked-string filter stable if we use the value $R = 1$? (If you want a hint, peek at the next problem.)

4. [Karplus and Strong, 1983], [Jaffe and Smith, 1983] We can estimate the time constant of each harmonic of the plucked-string filter in Section 5 as follows. A phasor at frequency ω is diminished in amplitude by the factor $|\cos(\omega T/2)|$ for every trip around the loop. After k samples, the phasor takes $k/(L + \frac{1}{2})$ round-trips. Define the *time constant* of a particular resonance to be the time in seconds that it takes for the amplitude of the resonance response to decrease by a factor $1/e$ (about 37 percent).

(a) Continuing the argument above, derive an approximate expression for the time constant of the nth harmonic of a plucked-string filter with loop delay L samples.

(b) Put the expression from Part (a) in terms of the actual frequency of the harmonic in Hz and the actual fundamental frequency of the plucked-string filter itself.

5. Rearrange Eq. 6.7 so that it uses only one multiplication per sample. Draw the corresponding signal flowgraph, analogous to Fig. 6.2.

6. Why is the low-frequency delay δ of an allpass filter always positive in practical situations?

7. Derive the formula for tuning the fundamental frequency of the plucked-string filter, Eq. 8.2, from Eq. 8.1.

8. That the particular one-pole, one-zero filter in Eq. 6.6 has a constant magnitude response is no miracle. It results from the fact that the order of the numerator coefficients is the reverse of those in the denominator. That is, the transfer function is of the form

$$\mathcal{H}(z) = \frac{a_0 z^{-n} + a_1 z^{-(n-1)} + \cdots + a_n}{a_0 + a_1 z^{-1} + \cdots + a_n z^{-n}}$$

Prove that all filters of this form are allpass.

9. [Fettweis, 1972] You might guess from Problem 8 that the one-pole, one-zero allpass filter is the first in a family of allpass filters that approximate constant delay at low frequencies. If you did guess that, you'd be right. The paper by Fettweis cited in the Notes gives an exceedingly elegant derivation of this family, based on a continued-fraction expansion. The next member has the transfer function:

$$\mathcal{H}(z) = \frac{z^{-2} + bz^{-1} + a}{1 + bz^{-1} + az^{-2}}$$

where

$$b = 2\,\frac{2 - \mu}{1 + \mu}$$

$$a = \frac{(2 - \mu)(1 - \mu)}{(2 + \mu)(1 + \mu)}$$

and μ is the desired delay in samples, playing the role that δ does in the one-pole, one-zero case.

(a) Prove that this filter is stable for $\mu > 1$. (In effect, we now have a built-in delay of one sample, and we should specify the delay in the range one to two samples.)

(b) Investigate the quality of the approximation to constant delay numerically, comparing it to the one-pole, one-zero filter. Decide when, if ever, it is a good idea to use it in the plucked-string instrument instead of the simpler filter.

10. If a two-pole, two-zero allpass filter has complex poles and zeros at angles $\pm\omega_c$, it will approximate linear phase in the range of frequencies corresponding to ω_c. Prove that and derive formulas for the phase delay analogous to Eqs. 7.7 and 7.8.

11. Suppose you replaced the lowpass filter in the plucked-string instrument with the highpass filter with equation

$$y_t = \tfrac{1}{2}[x_t - x_{t-1}]$$

What effect do you think this would have on its sound? Listen to it!

12. Karplus and Strong [1983] suggest that the following filter equation produces drumlike sounds:

$$y_t = \begin{cases} +\tfrac{1}{2}(y_{t-L} + y_{t-L-1}) & \text{with probability } b \\ -\tfrac{1}{2}(y_{t-L} + y_{t-L-1}) & \text{with probability } 1-b \end{cases}$$

where y_t is the output signal at time t.

(a) For what value of b does this reduce to something very close to the plucked-string algorithm? In what respect is it different?

(b) When $b = 0$ and the pitches are fairly high, Karplus and Strong describe the instrument as a "plucked bottle." Explain why this might be expected. (Hint: Recall Section 2.)

(c) Show that when $b = \tfrac{1}{2}$ it is approximately true that the mean-square value of the impulse response decays exponentially, and find the time constant.

13. Jaffe and Smith [1983] state

An effective means of simulating pick position is to introduce zeros uniformly distributed over the spectrum of the noise burst.

By "noise burst" they mean the initial input to the plucked-string filter. They go on to suggest filtering the initial noise input with an inverse comb filter of the form

$$w_t = x_t - x_{t-\gamma L}$$

where x_t is completely random (white noise), w_t is the newly prepared input signal, and γ is a number between zero and one representing the point on the string where it is plucked. Try to explain this, using what we learned about strings in Chapter 2. (Hint: A string plucked at a certain point will have certain harmonics missing.)

CHAPTER 7

Periodic Sounds

1 Coordinate systems

In one sense the simplest kinds of sounds are those that are periodic. Such signals can be represented by sums of phasors with frequencies that are integer multiples of the frequency of repetition, $1/T$, where T is the period.[†] The frequency of repetition is often called the *fundamental* frequency, and the multiples are called *harmonics*.

As we mentioned when we discussed the development of the plucked-string filter, just one periodic signal, played for a few seconds, will sound pretty boring, no matter what its fixed spectrum is. We don't get any really interesting sounds without some motion of the frequency content. But the study of the mathematics of periodic sounds, Fourier series, is a good place to start if we want insight into more complicated sounds. What's more, we'll find some ways to generate periodic sounds that can be used in more general situations.

We've seen several situations so far where we can think of signals as being composed of sums of sinusoids, or equivalently, phasors. When a signal is represented as such a sum, it's called a *frequency domain* representation of the signal. Actually, there are four different commonly used variants of frequency domain representations: Fourier series, the Discrete Fourier Transform (DFT), the *z*-transform, and the classical Fourier transform. But the intuition behind all of them is the same, and once you become familiar with one or two the others become easy. It's like picking up new computer languages once you learn your first.

The basic idea behind a frequency domain representation is really simple if you keep a geometric picture in mind: a vector v in ordinary three-dimensional space, as shown in Fig. 1.1. The vector v is written as a combination of the three unit vectors in each of the three coordinate directions:

[†] Don't confuse the period T of a continuous signal with the sampling interval T_s.

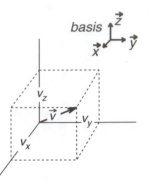

Fig. 1.1 A vector *v* in three-dimensional space.

$$v = v_x\vec{\mathbf{x}} + v_y\vec{\mathbf{y}} + v_z\vec{\mathbf{z}} \qquad (1.1)$$

This equation means that any vector v in our three-dimensional space can be obtained by adding three parts together, a vector in the x-direction of length v_x, a vector in the y-direction of length v_y, and a vector in the z-direction of length v_z. The component in the x-direction is thought of as a vector in the x-direction of length one, called $\vec{\mathbf{x}}$, times the number v_x, and similarly for the y- and z-components. The three unit-length vectors in the coordinate directions, $\vec{\mathbf{x}}$, $\vec{\mathbf{y}}$, and $\vec{\mathbf{z}}$, are called a *basis* for the three-dimensional space. I'll use the little arrows only for these unit-length basis vectors, just to emphasize their special meaning.

The numbers v_x, v_y, and v_z are called the *projections* of the vector v onto the respective basis elements. We'll denote the projection of one vector (say v) onto another (say w), by $\langle v, w \rangle$. Thus

$$v_x = \langle v, \vec{\mathbf{x}} \rangle$$
$$v_y = \langle v, \vec{\mathbf{y}} \rangle \qquad (1.2)$$
$$v_z = \langle v, \vec{\mathbf{z}} \rangle$$

In intuitive terms, v_x, the projection of the vector v in the x-direction, is the "amount" of v in that direction. If, for example, v is at right angles (orthogonal) to the x-axis, $v_x = \langle v, \vec{\mathbf{x}} \rangle = 0$.

When we get to Fourier transforms we're going to define several other examples of projection operators, but we're going to want them all to obey the same fundamental laws. For example, when we project the sum of two vectors onto a third, we want the result to be the sum of the individual projections. That is, suppose u, v, and w are any three vectors. Then we always want

$$\langle u + v, w \rangle = \langle u, w \rangle + \langle v, w \rangle \qquad (1.3)$$

This is a *distributive law*: projection distributes over addition. The projection $\langle u, v \rangle$ is also called the *inner product* of u and v. We'll use the terms interchangeably.

Also, when dealing with real-valued vectors we'll always want the projection operator to be symmetric; that is,

$$\langle u, v \rangle = \langle v, u \rangle \tag{1.4}$$

for all vectors u and v. (Later we'll be dealing with complex-valued vectors, and we'll have to modify this a bit.) Applying the symmetry law it's easy to see that the distributive law also works when the second vector is a sum:

$$\langle u, v + w \rangle = \langle u, v \rangle + \langle u, w \rangle \tag{1.5}$$

Because the three basis vectors we're using, $\vec{x}$, $\vec{y}$, and $\vec{z}$, are orthogonal to each other, they satisfy

$$\langle \vec{x}, \vec{y} \rangle = 0$$
$$\langle \vec{x}, \vec{z} \rangle = 0 \tag{1.6}$$
$$\langle \vec{y}, \vec{z} \rangle = 0$$

When the basis vectors are mutually orthogonal like this, we'll say the basis is an *orthogonal basis*.

We can now get a general idea of how the projection operator works by considering the projection $\langle v, w \rangle$ of any vector v on any other vector w. First, write v and w in terms of the their coordinates:

$$v = v_x \vec{x} + v_y \vec{y} + v_z \vec{z} \tag{1.7}$$

and

$$w = w_x \vec{x} + w_y \vec{y} + w_z \vec{z}$$

If we then form the inner product and use the distributive law in Eq. 1.3, the only terms that survive are the ones with like coordinates. The result is

$$\langle v, w \rangle = v_x w_x + v_y w_y + v_z w_z \tag{1.8}$$

This is a very important hint for getting the projection operator in other situations. Remember that it is the *sum of products of like coordinates, the sum being over all the coordinates.*

There's one wrinkle I need to point out before we can go on. The inner product of a vector with itself is, from the previous equation,

$$\langle v, v \rangle = v_x^2 + v_y^2 + v_z^2 \tag{1.9}$$

which is just the square of the length of the vector. However, this assumes the vector has real components. If a vector can have coordinate values that are complex numbers, which will often be the case in what follows, we still want an analogous statement to be true. For this reason we change the definition of inner product in Eq. 1.8 to

$$\langle v, w \rangle = v_x w_x^* + v_y w_y^* + v_z w_z^* \tag{1.10}$$

where as usual the ( )* denotes complex conjugate. This makes the inner product of a vector with itself

$$\langle v\,,v\rangle = |v_x|^2 + |v_y|^2 + |v_z|^2 \tag{1.11}$$

which is always real and non-negative. With this revision the formula for constructing an inner product is: *sum of products of like coordinates, the second complex-conjugated, the sum being over all the coordinates.*

These, then, are the two essential ingredients of what we call an *orthogonal coordinate system*:

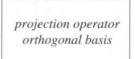

projection operator
orthogonal basis

With these simple geometric ideas we're going to derive all the mathematics we need for Fourier series, the Discrete Fourier Transform, and the z-transform. In each case it is just a matter of finding the appropriate basis and projection operator. Let's start with Fourier series.

2   Fourier series

We are now going to leap from three-dimensional space, in which geometric intuition operates comfortably, to a space that will at first appear strange: It will have an infinite number of dimensions. But the ideas in the previous section will work effortlessly. It's just a matter of being bold.

We want to represent periodic signals that are functions of a continuous time variable t, say for t in the range $0 \leq t \leq T$. In this chapter we'll always think of signals as repeating with period T outside this range. What inner product should we use? To answer this we need to decide what the coordinates of our space are; the rest will follow automatically. In fact, we have no choice. There is only one independent variable: t. As I just mentioned, this may seem strange, but there is really no reason we can't think of each particular value of t in the range 0 to T as a coordinate. The sum must then be an integral — the ''sum'' over the range of a continuous variable. The integral must be over the product of one function and the conjugate of the other, by the formula in the previous section, and this suggests the following definition for the inner product between periodic functions $f(t)$ and $g(t)$:

$$\langle f, g\rangle = \frac{1}{T}\int_0^T f(t)\,g^*(t)\,dt \tag{2.1}$$

in analogy to Eq. 1.10. Notice that we've remembered to take the complex conjugate of the second ''vector'' (really a function now). We've divided by T, the length of the interval, just so that the length of basis elements will turn out to be one; that's really just a matter of convenience. This inner product satisfies the distributive law, which you can verify easily because the integral of a sum is the sum of integrals.

What basis should we use? Well, the basis elements should be defined over this same range of continuous time t and should also be periodic. It wouldn't make sense to use any other kinds of elements to express such signals. The natural candidates for basis elements are the phasors with period T,

$$e^{jk\omega_0 t}, \quad k = \ldots, -1, 0, 1, 2, \ldots \tag{2.2}$$

where $\omega_0 = 2\pi/T$, the repetition rate in radians per sec. Notice that we've included the negative as well as the positive frequencies, which we always do when using phasors, so we can represent real functions like $2\cos\omega_0 t = e^{j\omega_0 t} + e^{-j\omega_0 t}$.

Before continuing, I want to point out something very important. The basis we're proposing is indexed by the integers. The functions we're going to represent with it are functions of the continuous variable t. In other words, the functions have coordinates that are indexed by the continuous index t. Therefore, when we arrive at the representation of $f(t)$ in terms of the basis in Eq. 2.2, we will have changed the coordinate system drastically — from one with a continuously indexed coordinate (time) to one with a discretely indexed coordinate (the phasor basis). At first this may seem impossible, but it works, although I'm sweeping some mathematical restrictions under the rug.

The next step in our routine is to check the orthogonality of the proposed basis. This is really very simple. Suppose we first consider two different basis elements, say $e^{jk\omega_0 t}$ and $e^{jm\omega_0 t}$, where $k \neq m$. The inner product in Eq. 2.1 yields an integrand we can evaluate immediately:

$$\frac{1}{T} \int_0^T e^{j(k-m)\omega_0 t} dt = \frac{e^{j2\pi(k-m)} - 1}{2\pi(k-m)} \tag{2.3}$$

where we have used the facts that $\omega_0 T = 2\pi$, and $k \neq m$. The complex exponential on the right is equal to one, because the exponent is an integral multiple of 2π, so the inner product is zero, as we wanted to demonstrate.

When $k = m$ the inner product in Eq. 2.3 becomes just the average value of unity, and the result is unity. Thus the proposed basis is not only orthogonal, but the length of each element is one.

We can now write any periodic function in terms of the basis:

$$f(t) = \sum_{k=-\infty}^{\infty} c_k e^{jk\omega_0 t} \tag{2.4}$$

in analogy to Eq. 1.7. Think of this as the periodic signal $f(t)$ expressed in a new coordinate system, with component c_k in the ''direction'' of the phasor $e^{jk\omega_0 t}$. That is, $f(t)$ is decomposed into a sum of phasors, and contains an amount c_k of the frequency $k\omega_0$. Equation 2.4 is called the *Fourier series* of $f(t)$, and we refer to the sequence c_k as the *spectrum* of the periodic signal $f(t)$.

How do we find the Fourier coefficient c_k? It is simply the projection of $f(t)$ on the kth basis element:

$$c_k = \langle f(t), e^{jk\omega_0 t} \rangle = \frac{1}{T} \int_0^T f(t) e^{-jk\omega_0 t} dt \tag{2.5}$$

the minus sign resulting from the complex conjugate operation for the second function in the inner product.

Notice one last thing. In the very common situation when $f(t)$ is real, there is the very simple relationship between c_k and c_{-k}:

$$c_k = c_{-k}^* \tag{2.6}$$

This follows because k appears only in the expression jk on the right-hand side of Eq. 2.5. Replacing k by $-k$ is therefore equivalent to replacing j by $-j$, which is the same thing as taking the complex conjugate. Thus, when $f(t)$ is real, the negative-k terms in the Fourier series are the complex-conjugates of the positive-k terms. Using this fact, and using Euler's formula for the phasor, we can rewrite the Fourier series in Eq. 2.4 as

$$f(t) = c_0 + 2 \sum_{k=1}^{\infty} \mathcal{Real} \{ c_k e^{jk\omega_0 t} \}$$

$$= c_0 + 2 \sum_{k=1}^{\infty} c_{real,\, k} \cos(k\omega_0 t) - 2 \sum_{k=1}^{\infty} c_{imag,\, k} \sin(k\omega_0 t) \tag{2.7}$$

where we have broken the Fourier coefficient c_k into its real and imaginary parts,

$$c_k = c_{real,\, k} + j c_{imag,\, k} \tag{2.8}$$

This is a convenient form when we want the Fourier series of a real signal in terms of sine and cosines instead of complex phasors.

We've just written down a fair amount of general material, so it's time for an example.

3 Fourier series of a square wave

The square wave shown in Fig. 3.1 is the time-honored first example of a Fourier series, and is interesting from both an aural and a mathematical point of view. We'll consider the simple case when the square wave alternates between the values $+1$ and -1, and remains at each value for $T/2$ sec, half the period. Equation 2.5 gives us the Fourier coefficient c_n in terms of the time function:

$$c_n = \frac{1}{T} \left[\int_0^{T/2} e^{-jn\omega_0 t} dt - \int_{T/2}^{T} e^{-jn\omega_0 t} dt \right] \tag{3.1}$$

Some straightforward algebra (good practice, see Problem 3) yields

$$c_n = \begin{cases} -2j/(n\pi) & n = 1, 3, 5, \ldots \\ 0 & \text{else} \end{cases} \tag{3.2}$$

from which we see that, for n odd,

$$\begin{aligned} c_{real,\, n} &= 0 \\ c_{imag,\, n} &= -2/(n\pi) \end{aligned} \tag{3.3}$$

The final Fourier series can then be written using Eq. 2.7:

$$f(t) = \sum_{n=1,3,5,\dots}^{\infty} \frac{4}{\pi n} \sin(n\omega_0 t) \tag{3.4}$$

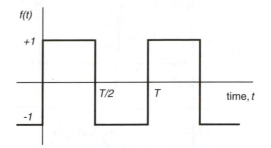

Fig. 3.1 Square wave of period T, taking on values ± 1 for half the period.

The first thing we might notice about this Fourier series is that it has only sine terms, no cosine terms. It's easy to see why: the original square wave is an odd function of t, and so is the sine function. That is, the function satisfies $f(-t) = -f(t)$; at any given negative value of t the function is the negative of what it is at the corresponding positive value of t. Any cosine terms would ruin this property. On the other hand, an even function of time, satisfying $f(t) = f(-t)$, can have only cosine terms in its Fourier series.

A second thing to notice is that the series has only odd harmonics. Again, this can be explained by considering symmetry. The original square wave satisfies $f(t) = -f(t+T/2)$. This is true of the odd harmonics of the sine, but not the even harmonics.

Having broken apart the square wave into its Fourier components, it's a good idea to verify that we can put the pieces back together. Figure 3.2 shows the sum of harmonics 1, 3, 5, 7, and 9. The result is recognizable as an approximation to a square wave, but there's only so much that five sine waves can do. The sum of the first twenty nonzero harmonics, shown in Fig. 3.3, is much more convincing. Notice that the approximation is worst at the sudden jumps, as you might expect.

The magnitude of the nth Fourier coefficient of the square wave is, from Eq. 3.2,

$$|c_n| = \begin{cases} 2/(n\pi) & n = 1, 3, 5, \dots \\ 0 & \text{else} \end{cases} \tag{3.5}$$

which is plotted in Fig. 3.4. This is the *spectrum* of the square wave: the amount of the phasor at the frequency $n\omega_0$. A quick look at this plot shows that the spectral content decreases rather slowly with the frequency. More precisely, the amount of the nth harmonic decreases as $1/n$ (skipping the absent even harmonics, of course). For example, $|c_{101}|$ is only about 2 percent smaller than $|c_{99}|$. This slow decay rate is closely associated both with how the square wave looks and how it sounds. A spectrum that falls off only as fast as $1/n$ is always associated with a signal that jumps

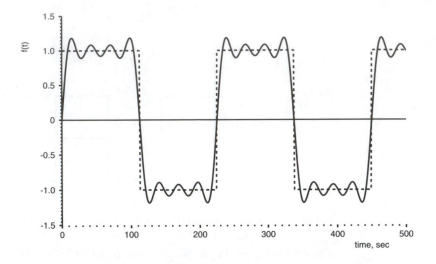

Fig. 3.2 The Fourier series for a square wave; terms up to the ninth harmonic are included. The ideal square wave is shown as a dashed line. The period $T = 225$.

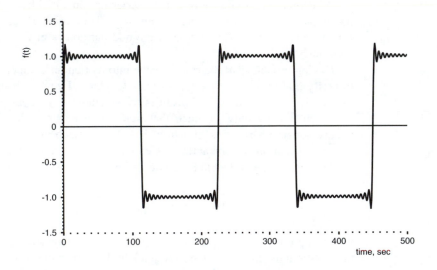

Fig. 3.3 The Fourier series for a square wave; terms up to the 39th harmonic are included.

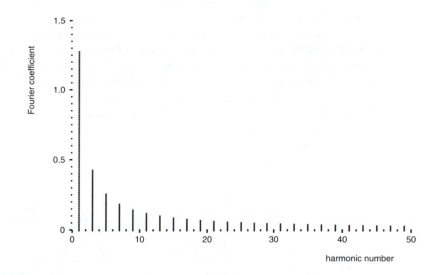

Fig. 3.4 The spectrum of a square wave; the magnitude $|c_n|$ of the coefficients.

suddenly, and, when it's periodic, sounds like a buzz saw. I'll elaborate on these important points in the next section.

Incidentally, the slow spectral decay that results from a discontinuity accounts for the fact that clicks plague computer music, and digital audio in general.[†] Segue carelessly from one stretch of sound to another, or do anything else without taking care to avoid a discontinuity, and you're sure to hear the click. Its $1/n$ spectrum is resplendent with high-frequency energy.

As we'll see shortly, a sharp pulse — which is just a discontinuity in one direction followed quickly by another in the opposite direction — has even more high-frequency energy than a single discontinuity, which is why the pops and scratches in phonograph records (remember them?) are so noticeable.

4 Spectral decay

The Fourier series for a function is the sum of an infinite number of terms, and it's not surprising that the rate at which the size of the terms falls to zero is critical. As we've just seen, the Fourier coefficients for the square wave in Fig. 3.4 converge to zero as $1/n$. What would be the general effect if the coefficients went to zero faster? Intuitively, we might suspect that the higher frequencies allow the signal to change faster, so we might guess that in general the faster the coefficients go to zero the smoother

---

[†] A phenomenon noted by F. R. Moore; see his book, referenced in the Notes to Chapter 1.

the function. Certainly it's true that the Fourier series is very smooth if only one or two harmonics are present.

An easy way to see the connection between the coefficient decay rate and smoothness is to remember that integration is a smoothing operation. Let's integrate the Fourier series in Eq. 3.4 term by term. The integral of the square wave itself is the triangle wave shown in Fig. 4.1. It is convenient to deal with a function that has average value zero, and this is easy enough to arrange by integrating from $T/4$ to t. That way the integral goes up to half the area of a half-period of the square wave, then down to minus that, and so on. I've used the word "smooth" loosely, but I can state quite precisely the sense in which the triangle wave is smoother than the square wave: the triangle wave is a continuous function of time, whereas the square wave is not. The square wave jumps between the values ± 1 in zero time.

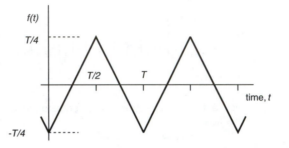

Fig. 4.1 A triangle wave; the result of integrating the square wave in Fig. 3.1.

The corresponding Fourier series of the triangle wave then becomes:

$$\int_{T/4}^{t} f(t)\,dt = -\sum_{n=1,3,5,\dots}^{\infty} \frac{2T}{\pi^2 n^2} \cos(n\omega_0 t) \qquad (4.1)$$

where we have substituted $\omega_0 = 2\pi/T$. The integration of the sinusoids $\sin(n\omega_0 t)$ has had the effect of multiplying the coefficients by a factor of $1/(n\omega_0)$, making the Fourier coefficients go to zero as $1/n^2$:

$$|c_n| = \begin{cases} T/(\pi^2 n^2) & n = 1,\ 3,\ 5,\ \dots \\ 0 & \text{else} \end{cases} \qquad (4.2)$$

As anticipated, the triangle wave, smoother than the square wave, has coefficients that go to zero much faster.

Figure 4.2 shows the result of putting together the Fourier series in Eq. 4.1 using only terms up to the ninth harmonic, in analogy to Fig. 3.2 for the square wave. We now do a much better job of approximation with these few terms than we did with the square wave, which makes sense because the omitted remaining terms in the series are smaller in magnitude.

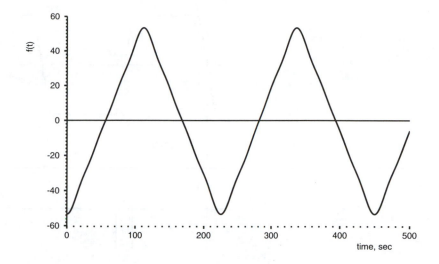

Fig. 4.2 The Fourier series for a triangle wave; terms up to the ninth harmonic are included.

5 Pulses

Differentiating a signal has an effect opposite to that of integrating, accentuating its fluctuations and multiplying its nth Fourier coefficient by n. Now let's turn the tables and differentiate the square wave in Fig. 3.1, instead of integrating it.

To be mathematically precise, the square wave's derivative is zero everywhere except at integer multiples of $T/2$, where it is undefined. It looks as if we can't deal with the derivative of a square wave at all. But things are not as bad as they seem, because we can think of the square wave as a limiting case of a smoother function, one that moves between -1 and $+1$ in a very short (but positive) time ε, rather than instantaneously; see Fig. 5.1(a). The derivative of this approximating function, shown in Fig. 5.1(b), is well defined: it is zero except for brief intervals of length ε, during which it is $\pm 2/\varepsilon$, the amount the function must change divided by the time it has to make that change.

Think of ε as very small compared to the period T of the signal; so small, in fact, that we can't hear any difference between the approximate version and a true square wave. For all practical purposes we can deal with the approximate square wave and its derivative instead of the corresponding ideal functions. The approximations are actually a better reflection of what exists in nature anyway, because signals can't jump from one value to another instantaneously in the real world.

To get the Fourier series of a signal like the one in Fig. 5.1(b), differentiate the Fourier series for a square wave, Eq. 3.4, yielding

$$f(t) = \frac{8}{T} \sum_{n=1,3,5,\ldots}^{\infty} \cos(n\omega_0 t) \tag{5.1}$$

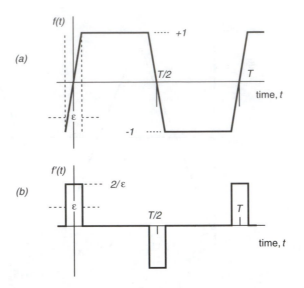

Fig. 5.1 (a) A function that approximates a square wave but is better behaved; (b) its derivative.

As usual, we used the fact that $\omega_0 = 2\pi/T$. We now have a very interesting situation. The differentiation multiplied the Fourier coefficients by n, resulting in a Fourier series where the odd-numbered coefficients don't decrease in magnitude with n at all. I've emphasized this by writing the coefficient $8/T$ outside the summation. This is not unreasonable: if we want to use a Fourier series to represent the sequence of pulses in Fig. 5.1(b) — a signal with sharp spikes — we might well expect the series to contain very high-frequency sinusoids with undiminished amplitudes. Figure 5.2 shows the result of using terms in Eq. 5.1 up to the 39th harmonic. We are trying to approximate a very wild function, so we can't expect to do nearly as well as when we tried to approximate the much tamer triangle wave in the previous section, which is two integrations smoother.

We can think of the derivative of a square wave, which is shown approximated in Fig. 5.1(b), as approaching a limiting function as ε approaches zero. The pulses become infinitely narrow, and their height becomes infinitely high, but in a controlled way: the area of each pulse stays fixed, in this case at $\varepsilon \times (2/\varepsilon) = 2$. These pulses are called δ *functions*, and are very convenient things to have around for mathematical manipulations, even though they aren't really functions in the ordinary sense of the word (see the Notes). The ideal derivative of the square wave is shown in Fig. 5.3. The only thing that matters about each pulse is its area and position, and we'll represent the ideal pulse of unit area positioned at $t = 0$ as $\delta(t)$. Thus the pulse train $p(t)$ in Fig. 5.3 can be written as

$$p(t) = \sum_{k=-\infty}^{\infty} (-1)^k 2\delta(t - kT/2) \tag{5.2}$$

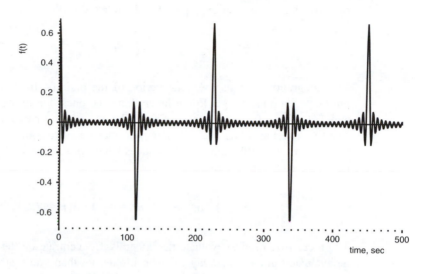

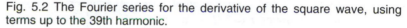

Fig. 5.2 The Fourier series for the derivative of the square wave, using terms up to the 39th harmonic.

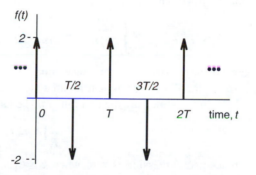

Fig. 5.3 The derivative of the square wave shown in Fig. 3.1. The arrows on the spikes indicate these are δ functions, and the time axis is continuous. The ordinate is the pulse area.

Each term $2\delta(t - kT/2)$ represents a pulse at $t = kT/2$ with area 2. We also need to multiply by $(-1)^k$ to take into account the fact that the pulses are alternately positive and negative.

You should keep two pictures of pulses in your head, for use in different contexts: first the realistic rectangular pulse approximations shown in Fig. 5.1(b); and second the δ functions shown in Fig. 5.3. The first kind are closer to signals that exist in the real world; the second allow us to do some slick mathematical manipulations.

To illustrate how useful the concept of a δ function is, suppose we started with the ideal pulse train $p(t)$ in Eq. 5.2, and wanted to get its Fourier series, reversing the line of derivation in this section. Apply Eq. 2.5 to get the kth Fourier coefficient c_k:

$$c_k = \frac{1}{T}\int_0^T p(t)\,e^{-jk\omega_0 t}\,dt \qquad (5.3)$$

The integration extends over one period of the periodic train of pulses $p(t)$, so the only contributions we get to the integral are the ones due to the δ functions at $t = 0$ and $T/2$.[†] The δ functions are zero everywhere except over two infinitesimal intervals. The first interval, which we'll denote by I, includes values of t near $t = 0$; the second interval, which we'll denote by J, includes values of t near $t = T/2$. The two integrals in Eq. 5.3 can then be written

$$c_k = \frac{2}{T}\int_I \delta(t)\,e^{-jk\omega_0 t}\,dt - \frac{2}{T}\int_J \delta(t - T/2)\,e^{-jk\omega_0 t}\,dt \qquad (5.4)$$

To proceed we need to evaluate the integrals. Concentrate on the second; the first will be evaluated in the same way. This is a lot easier than you might think at first. The δ function in the second integral is zero everywhere except over an infinitely narrow interval around $t = T/2$, so the only value of the phasor that could possibly matter is its value at that point. In fact, over that infinitesimal interval the phasor can be considered constant, equal to its value at $t = T/2$. Therefore, the second integral in Eq. 5.4 can be rewritten

$$e^{-jk\omega_0 T/2}\int_J \delta(t - T/2)\,dt = (-1)^k \int_J \delta(t - T/2)\,dt \qquad (5.5)$$

where we have used the fact (once more) that $\omega_0 = 2\pi/T$. The rest is easy; the integral of the δ function is just its area, 1, so the entire second integral is equal to $2(-1)^k$. In the same way the first integral is equal to 2, so Eq. 5.4 becomes

$$c_k = \frac{2}{T}\big[1 - (-1)^k\big] \qquad (5.6)$$

The bracketed expression is zero when k is even, and 2 when k is odd, so we finally get that c_k is zero when k is even, and $4/T$ when k is odd. Substituting these coefficients in the Fourier series Eq. 2.7 gets us right back to Eq. 5.1.

This may seem like a lot of work to get us back where we started, but the point is that we could have started with the ideal pulse train in Fig. 5.3 and obtained its Fourier series directly. This is just one example of how useful δ functions are. The important fact to remember about δ functions is that when we integrate over them, they "punch" out the value of the integrand:

[†] We're assuming the period begins immediately before $t = 0$ and ends immediately before $t = T$. This convention is arbitrary, but we should be consistent about it, always including two pulses in a single period. Perhaps a better way to write the formula for the Fourier coefficient is as an integral over one period.

$$\int_{-\infty}^{\infty} \delta(t-\tau)\,\psi(t)\,dt \;=\; \psi(\tau) \tag{5.7}$$

assuming of course that $\psi(t)$, the rest of the integrand, has a well defined value and is reasonably smooth at the value of t where the δ function spikes.

6 Continuous-time buzz

The signal we studied in the previous section, the derivative of a square wave, is composed of pulses that alternate in sign, and its Fourier series is missing even harmonics. The derivation that concluded with Eq. 5.6 shows that these two properties are closely related. A more standard sequence of pulses doesn't alternate in sign, and has every harmonic present. Intuitively, this is the harmonically richest signal we can generate with a given period T. Let's use our newly acquired δ functions to derive its Fourier series.

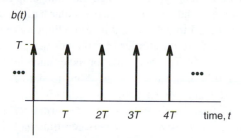

Fig. 6.1 Buzz, a sequence $b(t)$ of positive pulses of area T spaced T sec apart. The arrows on the spikes indicate these are δ functions, and that the time axis is continuous.

The signal we're interested in, which we'll call "buzz" and denote by $b(t)$, is shown in Fig. 6.1, a simple sequence of positive pulses. We haven't decided what the area of each pulse is, but it really doesn't matter, as long as it's the same from pulse to pulse. Take the area to be equal to the period T, which will make the Fourier coefficients one. We can then write the buzz signal in the form of a sum of δ functions, as in Eq. 5.2:

$$b(t) \;=\; \sum_{k=-\infty}^{\infty} T\delta(t - kT) \tag{6.1}$$

where we have included a factor representing the area, T, of each pulse. Its nth Fourier coefficient is $1/T$ times the integral over one period of the single δ function in that period, which is T. That is, every Fourier coefficient is one. Thus we have the Fourier series for $b(t)$:

$$b(t) \;=\; \sum_{n=-\infty}^{\infty} e^{jn\omega_0 t} \tag{6.2}$$

where as usual $\omega_0 = 2\pi/T$, the frequency in radians per sec. We can also write this as the following cosine series using Eq. 2.7:

$$b(t) = 1 + 2\sum_{n=1}^{\infty} \cos(n\omega_0 t) \qquad (6.3)$$

These last two forms of the Fourier series for the buzz signal reflect some fundamental facts about periodic waveforms and their spectra, and have a nice intuitive interpretation. They say:

If we add up phasors at all the harmonics of a given fundamental frequency 1/T Hz, all with the same amplitude, they cancel out to zero almost everywhere, except at pulses spaced every T sec, where they reinforce one another to produce δ functions.

Perhaps it's easiest to understand the way this works from the cosine series in Eq. 6.3. Imagine the sum of an infinite number of cosine waves, at all harmonics of the fundamental frequency, and all with the same amplitude. At any time that is not an integer multiple of the period T, any particular harmonic is as likely to be negative as positive. It's not hard to believe that in some limiting sense the sum will be zero. On the other hand, at integer multiples of T the cosines all have a positive value, and therefore add up to infinity. That accounts for the δ functions at integer multiples of T.

Here's another important thing to notice. The spectrum of buzz has harmonics spaced every $1/T$ Hz apart, as shown in Fig. 6.2. The $1/T$ is important: the larger the period T is, the closer together the harmonics are spaced. This is a simple manifestation of a pervasive duality between time and frequency. The more closely things are spaced in one domain, the more widely they are separated in the other.

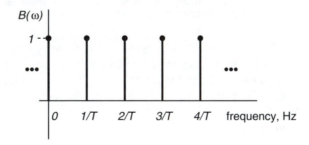

Fig. 6.2 The spectrum of the buzz signal.

7   Digital buzz

So far in this chapter we've assumed the time axis is continuous. The periodic signals we've expanded in Fourier series are functions of a continuous variable t. This implies that the harmonics can extend to infinity: we need an infinite number of harmonics to put together a train of ideal δ functions. But what about digital signals? We appear to be in something of a bind, because it doesn't make sense to sample the sequence of δ functions shown in Fig. 6.1. At any particular sampling time the ideal spikes have the value 0 or infinity! What we want is the discrete-time signal analogous to this sequence. This underscores the fact that continuous time, which implies the possibility of infinite frequency, is a mathematical idealization. Thinking of infinite Fourier series helps us understand nature, but in the last analysis, digital signals, which are nothing more than sequences of numbers, are more concrete and certainly closer to the representations in a computer. In digital audio, there aren't any frequencies above the Nyquist frequency.

The proper discrete-time counterpart of the continuous-time buzz shown in Fig. 6.1 is intuitively clear: It's the digital signal that takes on a positive value every P samples and is zero at other samples. In the frequency domain we expect it to be composed of all the harmonics up to the Nyquist frequency. This is true, but we need to fill in some details.

The starting point for understanding the discrete-time buzz signal is the finite sum of phasors:

$$\sum_{n=-N}^{N} e^{jn\omega t} \tag{7.1}$$

This looks like part of the Fourier series in Eq. 2.4 with equal-strength harmonics, but be careful! First, we want this to represent a discrete-time signal, so the time variable t is restricted to integer values. Second, ω is not defined as above in terms of the repetition period T of a continuous-time signal. We're going to choose ω to get the digital version of a buzz signal.

The sum in Eq. 7.1 is a geometric series, and we can derive the following closed form for it (see Problem 7):

$$\sum_{n=-N}^{N} e^{jn\omega t} = \begin{cases} \dfrac{\sin((2N+1)\omega t/2)}{\sin(\omega t/2)} & \text{if } \omega t \neq m2\pi \\ 2N+1 & \text{if } \omega t = m2\pi \end{cases} \tag{7.2}$$

where m is any integer. To use this for our purpose, we set the frequency ω to an integer fraction of the sampling rate; that is, set $\omega = 2\pi/P$ radians per sample, where P is an integer. We also choose the number of harmonics N so that the sum goes up to, but not past, the Nyquist frequency. This means that we choose $N\omega = 2\pi N/P$, the highest frequency in the sum, to be as close to π as possible without exceeding it. There are really two cases here: when P is even we choose $N = P/2$ and actually reach the Nyquist frequency in the sum; when P is odd, we choose $N = (P-1)/2$ and don't. I'll work out the latter case and leave the former case for Problem 8.

Proceeding then with the case when P is odd, we have $2N + 1 = P$, and the right-hand side of Eq. 7.2 simplifies considerably, because with the choices $\omega = 2\pi/P$ and $N = (P-1)/2$, the sine on the top, $\sin((2N+1)\omega t/2) = \sin(\pi t)$, becomes zero. (Remember that t is an integer.) Equation 7.2 can therefore be written

$$1 + 2\sum_{n=1}^{(P-1)/2} \cos(n2\pi t/P) = \begin{cases} 0 & \text{if } t \neq 0 \text{ mod } P \\ P & \text{if } t = 0 \text{ mod } P \end{cases} \tag{7.3}$$

This is exactly what we wanted: The sum of sinusoids using frequencies that are integer multiples of the sampling rate divided by P is precisely the digital buzz signal, with a nonzero value every P samples.

When P is even, things are slightly complicated by the fact that the sum goes up to and actually includes the Nyquist frequency. The result corresponding to Eq. 7.3 is

$$1 + 2\sum_{n=1}^{P/2-1} \cos(n2\pi t/P) + (-1)^t = \begin{cases} 0 & \text{if } t \neq 0 \bmod P \\ P & \text{if } t = 0 \bmod P \end{cases} \quad (7.4)$$

We can interpret the difference as follows: To make things work out to a buzz signal of the same form, we need to weight the term corresponding to the Nyquist frequency itself by one instead of two, just as we do for the zero frequency. The term $(-1)^t$ is that term, a phasor at the Nyquist frequency.

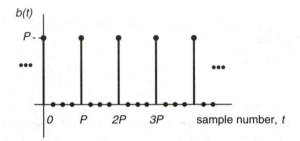

Fig. 7.1 Digital buzz. This shows samples of a discrete-time signal, not δ functions.

Figure 7.1 shows what the digital version of buzz looks like. It is certainly very easy to generate, since it's just a constant every P samples, with zeros in between. We now know that this represents exactly what we hoped it would: a combination of all possible harmonics of the fundamental frequency f_s/P (where $f_s = 1/T_s$ is the sampling frequency in Hz) up to the Nyquist frequency, equally weighted. Be sure you understand the distinction between the digital buzz signal illustrated in Fig. 7.1 and its analog counterpart in Fig. 6.1.

8 Synthesis by spectrum shaping

We're finally in a position to do something useful with our frequency representation. Suppose we want to produce a periodic sound with a given spectrum. The simplest and most obvious way is to add up the required sinusoids, each with the desired amplitude. This is called *additive synthesis*. Evaluating all those sinusoids and weighting them properly can be quite expensive computationally, as you can see by doing a little bookkeeping. Suppose the Nyquist frequency is 22 kHz, and the fundamental frequency of a desired periodic tone is 110 Hz. We then need to do the following for each sample: find the values of 200 different sinusoids, multiply them by the required weights, and then add them up. Usually, if we're worried about efficiency, we don't compute the values of sinusoids from scratch, but rather look them up in tables. Still, ignoring any possible table interpolation, we need to do 200 table lookups, 200 multiplications, and 199 additions for every output sample.

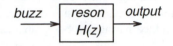

Fig. 8.1 Shaping the spectrum of buzz with a reson filter.

Filtering gives us a much faster way to get periodic signals with shaped spectra. The idea is very simple. Just pass a buzz signal through a filter, as shown in Fig. 8.1. Because we want to generate a digital signal, we'll use a digital buzz and a digital filter. Each harmonic in the buzz signal will be modified by the filter in a way specified by the filter's frequency response. As an example, suppose we use a two-pole resonator, the reson filter discussed in Chapter 5. The magnitude response $|H(\omega)|$ peaks at some center frequency and has the general shape illustrated in Fig. 8.2. Let's denote by ω_0 the periodic repetition frequency of the digital buzz signal in radians per sample. The nth harmonic of the buzz signal occurs at the frequency $n\omega_0$ radians per sample, and will have its magnitude multiplied by $|H(n\omega_0)|$, the value of $|H(\omega)|$ at the frequency $n\omega_0$ in radians per sample. The overall result, therefore, is a periodic waveform with an overall spectrum shape determined by the filter frequency response.

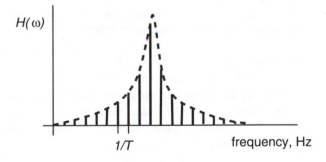

Fig. 8.2 Spectrum illustrating the shaping of digital buzz with a reson. The harmonics of the buzz signal are spaced $1/T$ Hz apart.

Let's look next at the amount of computation it takes to generate a signal this way. The buzz itself, as mentioned above, is trivial to generate by just using a nonzero value every P samples, as indicated in Eqs. 7.3 and 7.4. The filtering operation using a two-pole reson (see Eq. 4.3 in Chapter 5) requires only two additions and two or three multiplications (depending on whether there's a scale factor) per sample. So in the example considered at the beginning of this section, the filtering method requires only about one-hundredth as much computation per sample as the additive synthesis method.

There's another important reason why it's a good idea to shape the spectrum of buzz with a filter instead of adding together all the required harmonics. It's a reason that goes beyond issues of efficiency and begins to get at the real problems of generating sound on a computer. Return to the example above with a fundamental frequency of 110 Hz. How can we think about choosing 200 weights for each of the 200 harmonics? We don't have any tools to lean on for intuition.

We'd like something to turn over in our minds and manipulate. The picture of equally weighted harmonics being shaped by a filter, Fig. 8.2, gives us just such a structure. We can think about sliding the filter center frequency or adjusting its bandwidth on the fly — that is, while a particular stretch of sound is being generated (see Fig. 8.3).

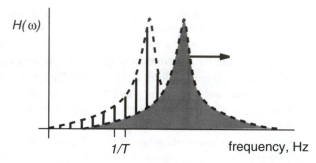

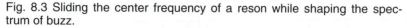

Fig. 8.3 Sliding the center frequency of a reson while shaping the spectrum of buzz.

As mentioned at the beginning of this chapter, perfectly periodic signals are not very interesting. The technique of spectrum shaping by filtering suggests interesting ways to change signals while keeping control of meaningful parameters like the bandwidth and center frequency of the overall spectrum shape. It's analogous to the way we deal with physical musical instruments, where we change sound quality by controlling a few easily grasped parameters, like string length, thickness, and tension.

We might also think of sliding the fundamental frequency of the buzz input. That's conceptually simple, but leads to an interesting technical problem, which I'll discuss in the next section.

9 Generating variable-frequency buzz

Suppose you want to change the frequency of buzz continuously, as it's generating samples. The problem that comes up is similar (but not identical) to the problem of tuning a plucked-string filter, which we discussed in Chapter 6. For example, suppose you are working at a sampling rate of 22,050 Hz and a period of $P = 20$ samples. To raise the frequency, you need to decrease the period. The smallest possible increase in frequency corresponds to a decrease in period from 20 to 19, and hence to a frequency increase of more than 5 percent. This hardly allows us to "slide" the frequency. After all, the interval of a semitone in the well-tempered scale corresponds to a ratio of $2^{1/12}$, or about 6 percent. What we hear if we restrict ourselves to integer periods is a strange scale rather than a glissando! But how can we create a pulse train with a period that is not an integer number of samples? In the world of digital signals, we specify signals only at sampling instants. If we wanted a period of 19.3 samples, how could we possibly place a pulse every 19.3 samples?

The solution is to return to Eq. 7.2, which we'll rewrite in terms of cosines as

$$1 + 2\sum_{n=1}^{N} \cos(n\omega t) = \begin{cases} \dfrac{\sin((2N+1)\omega t/2)}{\sin(\omega t/2)} & \text{if } \omega t \neq m2\pi \\ 2N+1 & \text{if } \omega t = m2\pi \end{cases} \tag{9.1}$$

When we last looked at this, we chose the frequency ω to be $2\pi/P$ radians per sample, an integer fraction of the sampling frequency. As we've just noticed, this severely restricts the range we can use when we want to move the frequency around.

Here's the point: There is no reason we can't use Eq. 9.1 when ω is any frequency whatsoever. This simple observation allows us to slide the frequency continuously, possibly changing it by some small amount every sample. Don't forget that since Section 7 we've been assuming t is the sample number, an integer.

Equation 9.1 is even more flexible: There is now no reason to choose the number of harmonics N so that all frequencies up to the Nyquist are included. We are perfectly free to choose any number of harmonics we want. Bear in mind, though, that if N is large enough to include frequencies *above* Nyquist, they will be aliased to frequencies below.

Implementing Eq. 9.1 is not trivial. For arbitrary ω there will inevitably be sample numbers t where ωt is close to, but not exactly equal to, an integer multiple of 2π. At those points the top expression will divide one very small number by another. If we don't take the proper precautions, that will lead to unacceptable numerical errors. I'll let you worry about that in Problem 9.

In the next chapter, we'll see how to translate the frequency domain ideas in this chapter to a numerical algorithm that can be applied directly to digital signals.

Notes

It is a curious fact that if you add up more and more terms of a Fourier series, as illustrated in Figs. 3.2 and 3.3, the overshoot precisely at a discontinuity never completely goes away. As the number of terms goes to infinity, the sum overshoots in the limit by about 8.95 percent times the size of the discontinuity. This is called the *Gibbs phenomenon*, after J. Willard Gibbs, who described it in *Nature Magazine* at the surprisingly late date of 1899. But Carslaw points out that "Wilbraham had noticed its occurrence . . ." in 1848. See the classic

    H. S. Carslaw, *Introduction to the Theory of Fourier Series and Integrals*, (third revised edition), Dover, New York, N.Y., 1930.

The mathematically ideal pulse called the δ function was introduced by the British Physicist Paul Adrien Maurice Dirac (1902–1984), who used it in quantum mechanics. In fact, it is sometimes called the "Dirac δ function." However, the choice of the letter "δ" for a function that is nonzero only under special circumstances appears to come not from Dirac's name, but from the Kronecker δ, which predates Dirac. The Kronecker δ is the integer-valued function δ_{ij} on integers i and j that is defined to be one if $i = j$ and zero otherwise.

It took many years for the δ function to gain mathematical respectability. For example, a standard midcentury applied mathematics book (*The Mathematics of Physics and Chemistry*, by H. Margenau and G. M. Murphy, Van Nostrand, New York, N. Y., 1943) states, "From a mathematical point of view such a function is a monstrosity. . . ." Monstrosity or not, the δ function has been given a firm mathematical foundation and is now indispensable. For more, see

> A. Papoulis, *The Fourier Integral and its Applications,* McGraw-Hill, New York, N. Y., 1962.

In computer-music circles, generating sound by filtering a harmonically rich signal, as we discussed in Section 8, is called *subtractive* or *formant synthesis* — as opposed to *additive synthesis*, where we add up each sinusoidal component.

The use of Eq. 9.1 for generating a buzz signal with continuously variable frequency was suggested in

> G. C. Winham and K. Steiglitz, "Input Generators for Digital Sound Synthesis," (Letter to the Editor), *J. Acoust. Soc. Amer.*, vol. 47, no. 2, (Part 2), pp. 665–666, Feb. 1970.

As described there, the fast way to implement Eq. 9.1 is to look up sinusoid values in a sufficiently finely sampled, precomputed table. This is an example of the general technique called *wavetable synthesis*.

Problems

1. Show that for the usual geometric coordinate system in two-dimensional space (the Euclidean plane), the inner product $\langle v, w \rangle$ in Eq. 1.8 is

$$\langle v, w \rangle = |v| \cdot |w| \cos\theta_{vw}$$

where θ_{vw} is the angle between the vectors v and w.

2. Repeat for three dimensions.

3. Verify Eq. 3.2, the formula for the kth Fourier coefficient of a symmetrical square wave.

4. Generalize the observations in Section 3 about the relationships between a waveform's symmetry on the one hand and the presence or absence in its Fourier series of sines or cosines, or even or odd harmonics.

5. Find the Fourier series for the following continuous-time function, periodic with period T sec:

$$f(t) = \begin{cases} 1 & \text{if } 0 < t < \alpha \\ 0 & \text{if } \alpha < t < T \end{cases}$$

where $0 < \alpha < T$. This is a pulse of width α. Check your result against what we know about the symmetrical square wave for the case $\alpha = T/2$.

6. Study the magnitude of the Fourier coefficients of the signal in Problem 5 as α varies from 0 to T. Plot out several cases and draw some conclusions.

7. Derive the closed form for the finite sum of equally weighted harmonics, Eq. 7.2.

8. Check the derivation of Eq. 7.4, the discrete-time buzz signal for even period P.

9. Implement a buzz generator based on Eq. 9.1, and test it by generating glissandos followed by bandpass filters. Use wavetable synthesis for speed, as suggested in the Notes. The main problem here is that numerical problems occur when ωt is close to an integer multiple of 2π. It's your job to decide how close is "close," and to do the right thing in those cases.

10. Extend Eq. 1.4, which expresses the symmetry of the inner product, to the case when the vectors have coordinate values that are complex numbers.

The Discrete Fourier Transform and FFT

1 Circular domains

We're now in a position to enjoy one of the great triumphs of signal processing: the direct numerical calculation of the frequency content of real digital signals, at lightning speed. The algorithm we develop, the *Fast Fourier Transform* (FFT) is one of the most widely used in all of science. I've already used it without telling you — for example, to compute the spectrograms of real and synthetic plucked-string notes in Chapter 6.

Here's the plan for this chapter. First, we'll derive the mathematical transformation that the FFT computes, called the *Discrete Fourier Transform* (DFT). We'll use the same geometric approach that we used in Chapter 7 to derive Fourier series. The DFT will have its own special basis and projection operator, just like the Fourier series representation. Then we'll see how it can be computed efficiently.

Be careful to distinguish between the DFT, which is an abstract mathematical transformation independent of any questions of efficiency, and the FFT, which is an efficient *algorithm* for computing it. The FFT is in such common use that this distinction is often blurred.

The starting point for the DFT is very close to reality: it's just a file of numbers, representing a finite sequence of N samples of a digital signal. This is not the case when we use a Fourier series representation, where we begin with a time function defined for *all* real values of t in the interval $[0, T]$.[†] No file is large enough to hold all that information.

Now I want to emphasize a point that may appear paradoxical, but is actually perfectly logical. It doesn't matter whether we think of a signal as *defined only on a finite*

[†] Just a reminder: We continue to denote the period of a continuous signal by T, and the sampling interval by T_s.

interval or as *periodic*. For example, when dealing with Fourier series in the previous chapter, we always thought of a periodic signal as repeating its basic period forever, from the infinite past to the infinite future. However, once a periodic signal is determined for all the values of time t within any basic interval of length T, it is determined at all other values of t. Mathematically we can think of the domain as either the finite line segment or the infinite line composed of repeated segments. If the signal is represented by a Fourier series in the base interval, it will also be represented that way in repeated segments outside that interval, because the sinusoids in the Fourier series are all periodic with period T.

Notice that there is an important difference between a signal that is defined only on a finite interval and is thought of as repeating outside that interval, and one that is defined for all time but is zero outside the interval. It certainly makes a big difference when you listen to sound. A signal defined for all time that is chopped off so that it is zero outside the interval $[0, T]$ sounds quite different from the version that repeats periodically. We'll see in Chapter 10 that the chopped signal has a very different spectrum.

A good way to visualize a periodic signal is to wrap it around a circle — that is, to make its domain circular. In fact, this is how we've been viewing the frequency response of digital filters. You can think of the frequency response as being defined only in the strip of frequencies between $-f_s/2$ and $+f_s/2$. Or, equivalently, you can think of it as repeating forever with period f_s because it's defined on the unit circle in the z-plane. What's important is the fact that the entire frequency response is determined by its values in one basic strip of length f_s. In the rest of this chapter we'll be using circular domains for both the time domain of discrete-time signals, and the frequency domain of their frequency content defined at discrete points. For example, Fig. 1.1 shows the two ways of thinking of an eight-point segment of a digital signal x.

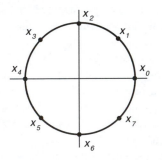

Fig. 1.1 A finite (eight-point) segment of a digital signal *x*, shown defined on a linear domain (top) and, equivalently, a circular domain (bottom).

Notice that we number the points starting from zero, so the signal points are x_0 through x_7. This is convenient when we think of the signal as periodic, because $\cdots x_{i-8} = x_i = x_{i+8} = x_{i+16} \cdots$. Using modular notation,

$$x_i = x_{i \bmod N} \tag{1.1}$$

for any i, where N is the period.

2 Discrete Fourier Transform (DFT) representation

We start with a finite, length-N segment of a digital signal, $x_0, x_1, \ldots, x_{N-1}$; something quite tangible that we typically capture on a computer and store in a file. The machinery for developing its frequency-domain representation is already in place. All we need to do is turn the key.

As before, we begin by choosing the inner product between two signals, say x and y, both defined for the integer time variable t. In the case of Fourier series, where the time variable is continuous, we use the integral of the product between the x and the complex conjugate y^*; now, when the time variable is discrete, we use the sum:

$$\langle x, y \rangle = \sum_{t=0}^{N-1} x_t y_t^* \tag{2.1}$$

You might expect that we divide by N, in analogy to the case of Fourier series, where we divided by the length of the time interval. The consequence of not dividing by N is that our natural basis elements will not have unit length, but that isn't critical. What's important is that they be orthogonal, which is unaffected. Actually, it comes down to whether the forward or the inverse transform has a factor in front, which is just a matter of convention.

The next step is to choose a basis. For a frequency-domain representation, we want a set of phasors that can represent any length-N segment of a digital signal. Before going further we need to execute a small maneuver to ensure that we end up with the universally accepted standard form of the DFT. Up to this point we've taken the basic frequency interval of a digital signal to be $[-\omega_s/2, +\omega_s/2]$. Now we're going to switch to the frequency interval $[0, \omega_s]$. You realize, of course, from the discussion in Section 1 of this chapter as well as in Chapter 3, that this change makes no difference whatsoever. Frequencies of discrete-time signals *are equivalent modulo the sampling frequency*. So, for example, the frequency $0.7\omega_s$ is equivalent to $-0.3\omega_s$. Just imagine a phasor jumping each sample by $0.7 \times 2\pi$ radians in the positive direction instead of by $0.3 \times 2\pi$ radians in the negative direction.

Return now to the problem of choosing a basis. We're dealing with sampled signals, so when we expand them in terms of phasors, we need to consider phasor frequencies only in the range $[0, \omega_s]$. The first temptation might be to choose as a basis *all* the phasors with frequencies in this range:

$$\{e^{jt\omega}\} \quad \text{for } 0 \le \omega < \omega_s \tag{2.2}$$

Notice that we don't go all the way up to the right end of the interval of possible frequencies, because for a digital signal, the frequency ω_s is the same as zero frequency.

This basis has a lot of elements in it — one for every real point between zero and ω_s. It turns out that this is overkill. We can get away with only N distinct frequencies in this range.

We choose the N frequency points equally spaced in the desired range from zero to the sampling frequency, namely 0, $2\pi/N$, $2(2\pi/N)$, ... , $(N-1)(2\pi/N)$ radians per sample. The corresponding basis is

$$\left\{ e^{jtk2\pi/N} \right\} \quad \text{for } 0 \leq k \leq N-1 \tag{2.3}$$

Think of the kth phasor in the basis as having the discrete time variable t and frequency $k2\pi/N$ radians per sample.

Checking the orthogonality of the basis elements is a calculation that may look familiar, a finite geometric series (see Problem 7, Chapter 7). We need to find the inner products

$$\left\langle e^{jtm2\pi/N}, e^{jtn2\pi/N} \right\rangle = \sum_{t=0}^{N-1} e^{jt(m-n)2\pi/N} \tag{2.4}$$

When $m = n$ the right-hand side is N, just the sum of N ones. I ask you to check that the inner product is zero when $m \neq n$ in Problem 1.

We have an inner product and a basis, so we are done. The frequency domain representation of the signal x is simply a sum of phasor basis elements, each weighted by the frequency content of the signal at the phasor's frequency:

$$x_t = \frac{1}{N} \sum_{k=0}^{N-1} X_k e^{jtk2\pi/N} \tag{2.5}$$

We use the uppercase X to represent the frequency content of the signal x, represented by a lowercase letter, a common convention. The factor $1/N$ is introduced here to avoid a factor in the forward transform, as mentioned above. The *frequency content*, or *spectrum* of x at frequency $k2\pi/N$ is obtained by the projection of x onto the kth basis element:

$$X_k = \left\langle x, e^{jtk2\pi/N} \right\rangle = \sum_{t=0}^{N-1} x_t e^{-jtk2\pi/N} \tag{2.6}$$

The minus sign comes from the complex conjugate in the inner product, Eq. 2.1. Since this tells us how to get the frequency content from the signal, we call Eq. 2.6 the *forward* DFT. The representation Eq. 2.5 tells us how to get the signal values from the frequency content, so therefore it is the *inverse* DFT.

In Problem 2 I ask you to check that using the factor $1/N$ in the inverse DFT is consistent with using no factor in the forward DFT.

We can also view the DFT rather abstractly as a way to convert one sequence of N complex numbers to another. The transformation is particularly well expressed in matrix-vector notation. Define x and X to be N-dimensional vectors with components x_t and X_k, respectively, and the $N{\times}N$ matrix F by

$$[F]_{kt} = e^{-j2\pi kt/N} \tag{2.7}$$

The notation indicates that the right-hand side is the element in row k and column t of the matrix F. The DFT in Eq. 2.6 can then be written very compactly as

$$X = Fx \qquad (2.8)$$

The inverse DFT results from inverting this matrix transformation by multiplying by the inverse of the matrix F:

$$x = F^{-1}X \qquad (2.9)$$

That is, multiplication by the matrix F transforms the signal into its spectrum, or frequency content. Multiplication by F^{-1} reverses the transformation, transforming the spectrum back to the signal. By the way, the inverse of the matrix F always exists; otherwise, the inverse DFT wouldn't exist (see Problem 3). This matrix-vector point of view, where the DFT is equivalent to matrix multiplication, is sometimes very useful in thinking about the DFT and its inverse.

3 The discrete frequency domain

When the DFTs of signals are computed, the result appears as a sequence of N complex numbers, representing the frequency content, or spectrum, of the original length-N signal segment. Let's make doubly sure you understand exactly how to interpret the result.

When we derived the DFT, it was mathematically convenient to take the frequency range to be $[0, \omega_s]$ instead of $[-\omega_s/2, +\omega_s/2]$. I made a point of the fact that these two ranges are completely equivalent; the latter interpretation simply means subtracting the sampling rate from all frequencies above the Nyquist frequency. Let's use $N = 8$ for illustration. ($N = 1024$ or 2048 is more representative of practical applications.) The DFT uses discrete frequency points numbered

$$0, 1, 2, 3, 4, 5, 6, 7 \qquad (3.1)$$

so this means interpreting the result by subtracting $N = 8$ from all indices above $N/2 = 4$, which yields

$$0, 1, 2, 3, 4, -3, -2, -1 \qquad (3.2)$$

A good way to visualize this is to draw the frequency points on a circle, as shown in Fig. 3.1.

Since N corresponds to the sampling rate, we need to divide by N to get the frequencies in terms of fractions of the sampling rate. Thus, the points in Eq. 3.2 correspond to

$$0, \frac{1}{8}, \frac{2}{8}, \frac{3}{8}, \frac{4}{8}, -\frac{3}{8}, -\frac{2}{8}, -\frac{1}{8} \quad \text{fractions of sampling rate} \qquad (3.3)$$

To take a realistic example, suppose we compute the 1024-point DFT of a signal sampled at 22,050 Hz. Points 0 to 512 correspond to consecutive multiples of the frequency 22,050/1024, starting with 0. Point 513 corresponds to minus the frequency of point 511; point 514 to minus the frequency of point 510, and so forth, down to point 1023, which corresponds to minus the frequency of point 1. This is illustrated in Table 3.1.

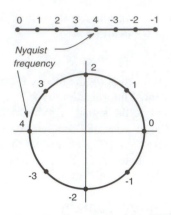

Fig. 3.1 The frequency domain of an eight-point DFT, shown defined on a linear domain (top), and, equivalently, a circular domain (bottom).

Point no.	Frequency, Hz
0	0.0
1	21.5
2	43.1
...	...
510	10,981.9
511	11,003.5
512	11,025.0
513	−11,003.5
514	−10,981.9
...	...
1022	−43.1
1023	−21.5

Table 3.1 The frequencies corresponding to the output points of a 1024-point DFT when the sampling rate is 22,050 Hz.

Next let's look at the very common case when the signal x_t is real-valued. In that case there's a very simple relationship between the frequency content at frequencies ω and $-\omega$. Remember that, from the discussion above, frequency indices are equivalent modulo N; if the frequency ω corresponds to point k in the DFT, frequency $-\omega$ corresponds to point $N-k$. Therefore, to look at the transform at frequency $-\omega$, examine point $N-k$ of the DFT of a length-N signal x_t, using the formula for the DFT, Eq. 2.6:

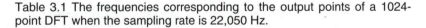

$$X_{N-k} = \sum_{t=0}^{N-1} x_t e^{-jt(N-k)2\pi/N} = \sum_{t=0}^{N-1} x_t e^{+jtk2\pi/N} \qquad (3.4)$$

Since we've assumed the signal values x_t are real-valued, this last expression is simply the DFT at point k, but with j replaced by $-j$. In other words, when x_t is a real-valued signal,

$$X_{N-k} = X_k^*$$ (3.5)

Thus, the values of the transform of a real signal at points from $(N/2+1)$ to $(N-1)$ are very simply related to the values at points from 1 to $(N/2-1)$: the value of the transform at point $(N-k)$ has the same magnitude as the value at point k, and the negative of its phase angle. For this reason, we plot the DFT of a real-valued signal only for points in the range 0 to $N/2$, corresponding to the range of frequencies from 0 to the Nyquist frequency.

Here's a very simple example of the transform of a real-valued signal. Suppose x_t is a 128-point segment of a cosine wave at the frequency $22\pi/128$ radians per sample:

$$x_t = 2\cos(22\pi t/128), \quad t = 0, 1, \ldots, 127$$ (3.6)

The signal can be rewritten using Euler's formula as

$$x_t = e^{jt22\pi/128} + e^{-jt22\pi/128}$$ (3.7)

The frequency can be written $11\times(2\pi/N)$, since $N = 128$, so this is just the sum of two basis elements, one at point 11, and the other at point -11, which is the same as point $128 - 11 = 117$. Knowing the signal is real-valued, we plot the DFT only up to the Nyquist frequency, point 64, and the signal's frequency shows up as a single unit spike at point 11 of the frequency plot, or 11/128 in fractions of the sampling rate.

In one sense, N samples of a real-valued signal contain only half the information contained in N samples of a complex-value signal. We've just shown that the DFT of a real-valued signal has (almost exactly) half the information as the DFT of a complex-valued signal, since the points above point $N/2$ are completely determined by the points below. This makes sense, and is evidence that, intuitively, the DFT preserves information. The important thing to note for future work is that the magnitude of the transform of a real-valued signal is an even function of frequency, and needs to be plotted only for positive frequencies, from zero to the Nyquist frequency.

4 Measuring algorithm speed

The great significance of the DFT is that it is a frequency transform that can be computed directly, with no approximations. It is a rare example of a mathematical tool that can be translated into action perfectly. What's more, it can be computed with surprising efficiency. The principle that makes this possible, *divide-and-conquer*, is interesting in its own right, and useful in other fields. Before we describe it, we need to discuss how we are going to estimate the time taken by an algorithm. That way, we will be in a position to compare different approaches to the same problem. Computer science students know all this; they have my permission to skip this and the next section.

Suppose we consider as an example the computation of the DFT in the most naive, straightforward way. The definition, Eq. 2.6, is a sum of N products for each of N frequency points. It appears that it requires two nested loops, each with limits zero to

$N-1$, and therefore a total of N^2 multiplications, and about the same number of additions. If we want to estimate the amount of time that would be taken by the algorithm, we could add the time for all the multiplications and additions. Thus, the transform of a 1024-point signal would seem to require about a million multiplications and about a million additions. We might even want to take into account the time required to compute the complex exponentials used as the basis elements, the array accesses (assuming the signal and the basis are stored in arrays), and maybe even the time required to read the data in and out.

Even this simple example shows that keeping precise track of all the time-consuming steps in an algorithm entails a lot of bookkeeping. Not only is this tedious, it tends to obscure the important trends. What we really want to know is that, roughly, the literal evaluation of the sums in Eq. 2.6 requires a number of operations roughly proportional to N^2. The reason for this is simple: Nothing is done more than N^2 times.

It turns out that we do not go far astray if we simply count multiplications and ignore everything else. That will give us the trend, which will be a good guide to algorithms that are efficient in practice. As you'll see shortly, we will achieve savings in algorithm speed of factors of thousands.

I've been intentionally vague here, because I don't want to get bogged down in a full-blown discussion of algorithm efficiency. But it's good to know that the notion of "trend" can be put on a solid mathematical basis. The terminology is that our naive algorithm takes $O(N^2)$ steps, read "order N^2" steps. We also say that the *asymptotic* time required for the algorithm is $O(N^2)$. The idea is that as N gets arbitrarily large, the number of steps is bounded by some constant times N^2. We will get by with the simple strategy of counting the number of times a representative operation is done.

To consider another example, how much time does it take to read in the data for an N-point DFT? We can answer this in "big-oh" notation without knowing exactly what happens during data input. Whatever it is that does happen, it happens once per data point. The time for input must be proportional to N, and we say that input takes $O(N)$ time.

A very simple argument supports this seemingly sloppy approach. Suppose we have two competing algorithms for the same task, the first taking $O(N)$ and the second $O(N^2)$. It's easy to see that if N is sufficiently large, the first will *always* win, regardless of the constants of proportionality we have ignored in such a cavalier manner. If, for example, the first algorithm takes exactly $1000N$ steps, and the second takes exactly N^2 steps, there is a break-even point determined by setting these two quantities equal to each other:

$$1000N = N^2 \tag{4.1}$$

The break-even point is therefore $N = 1000$, as shown in Fig. 4.1. Beyond that value of N the $O(N)$ algorithm is faster, even though it has a constant of proportionality a thousand times larger than the $O(N^2)$ algorithm. For 2000 points, for example, the linear-time algorithm is twice as fast.

In the case of the DFT, we win on all counts. Not only is there an algorithm that is asymptotically faster, the break-even value of N is quite small.

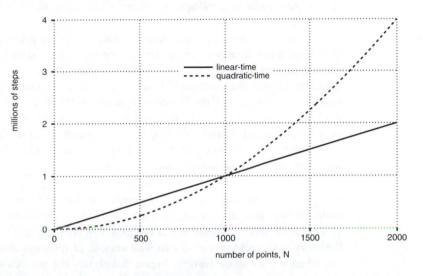

Fig. 4.1 A linear-time algorithm eventually beats any quadratic-time algorithm, regardless of the constants of proportionalities.

5 Divide-and-conquer

Divide-and-conquer, the principle we use to construct a fast algorithm for the DFT, is a basic tool of computer science. The simplest way to explain it is to show how it is used to find a good algorithm for sorting numbers.

Suppose we are given a list of N numbers, in arbitrary order, and we need to put them in a final array, sorted from smallest to largest. The naive approach, analogous to the naive way of computing the DFT, is to search for the smallest item, put it in the first location of the final array, search the remaining list for the next smallest, put that in the second location of the final array, and so on, as shown in Fig. 5.1. Each search takes a number of steps no more than the length of the list, which is no more than N, and we need to do the searches N times to fill up the final array. The naive sorting algorithm takes $O(N^2)$ time.

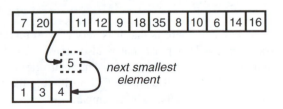

Fig. 5.1 Sorting the hard way, one at a time.

If we used the naive algorithm to sort a list of a million items, we would be in trouble. A million is not a lot of instructions for a computer to execute, but a million million is. At the rate of a million operations a second, a million million operations takes more than eleven days.

The basic problem with the $O(N^2)$ sorting algorithm is that it handles one number at a time. Each time we search the list for the next smallest, we are starting from scratch and not using the results of previous comparisons between numbers. A better idea is to process the numbers in batches. We can do this as follows: Sort the first two numbers on the list, say the numbers in positions 0 and 1. That is, put the smaller of the first two numbers in position 0, and the larger in position 1. Then sort the numbers in positions 2 and 3, and so forth, up to the numbers in positions $N-2$ and $N-1$, assuming for convenience that N is even. We can think of the result of this first pass as $N/2$ sorted lists containing two numbers each.

The next stage is to combine the $N/2$ pairs of two-element lists into sorted four-element lists. We then proceed by combining the $N/4$ four-element lists into $N/8$ eight-element lists, and so on, until we get to one sorted list of N elements. It's convenient in all this to assume that N is a power of two, but if it's not, we can easily pad the array to the next power of two with very large numbers and ignore the end of the list when we are done sorting. Figure 5.2 shows the sublists generated in the three stages when sorting eight numbers. This method of sorting is called, naturally enough, *merge sort*.

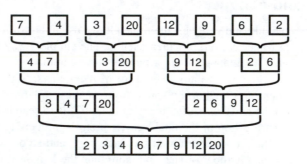

Fig. 5.2 Merge sort illustrated for eight elements. The top row contains eight lists of one element each. Each successive row contains half as many lists, each with double the number of elements.

A key point here is that it is very easy to merge a pair of sublists that are already sorted into a longer list that is also sorted. The basic idea is illustrated in Fig. 5.3. Start by comparing the smallest in each original sublist, say x and y. Suppose the smaller is x. Promote x to the first position in the new list, and move over to the next element in x's sublist. Next compare the current heads of the sublists. Promote the winner, move over in its sublist, and repeat, until you run out of elements in one of the sublists. Then just tack the end of the remaining sublist onto the final list.

How much work is required to merge two sublists? Estimate the work by counting the number of comparisons — certainly the number of operations is no worse than

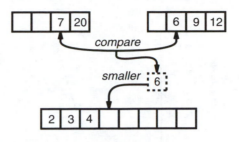

Fig. 5.3 Merging two sorted lists to obtain a longer sorted list.

proportional to that count. Now, each comparison results in another element being promoted to the new list. We may get a bonus at the end, if one of the sublists runs out of elements early, but in any event the number of comparisons cannot exceed the length of the final list, the total length of the two original sublists. In big-oh parlance, merging sorted lists is a *linear-time operation*, $O(N)$, where N is the total length of the lists involved.

Finally, let's estimate how many comparisons are involved in a merge sort. It's not hard to see that each merging stage, in which the size of the lists is doubled, takes no more than N comparison operations — because the total lengths of the lists generated is N and, by the argument in the previous paragraph, merging is linear-time. How many stages are there? That's easy. It's the number of times we have to double the size of lists from one to get past N, which is about $\log_2 N$. The total number of comparisons is therefore about $N\log_2 N$, and the remaining work is no worse than proportional to that. Merge sort, our new batch-processing sorting algorithm, therefore takes $O(N\log_2 N)$ time. Since logarithms to different bases differ only by constant factors, we'll drop the base-2 and write $O(N\log N)$ when we use big-oh notation, where constant factors don't matter.

The speedup in going from $O(N^2)$ to $O(N\log N)$ may seem unimpressive at first, but it is a breakthrough. It often allows us to do things that are otherwise impossible. Consider the simple example of sorting a million numbers. The logarithm (base-2) of a million is about 20, so merge sort takes no more than about twenty-million comparisons. As we mentioned, computers today can do a million things very fast, in less than a second. The factor of twenty means waiting twenty seconds, which is tolerable. But $N^2 = 10^{12}$, a million million — eleven days instead of twenty seconds!

The logarithm grows so slowly, in fact, that we can think of $O(N\log N)$ as being more like $O(N)$ than $O(N^2)$, as shown in Fig. 5.4. Doubling N just increases $\log_2 N$ by one, and the percent change in $N\log_2 N$ when N doubles becomes more and more like 100 percent (double) as N gets larger and larger.

The way we've described merge sort is nonrecursive; we start with the smallest tasks, pairwise merging N lists of length one, and work our way up to the final stage, where we merge two lists of length $N/2$. This is a perfectly reasonable way to think about how the algorithm works, and when translated to code leads to an efficient

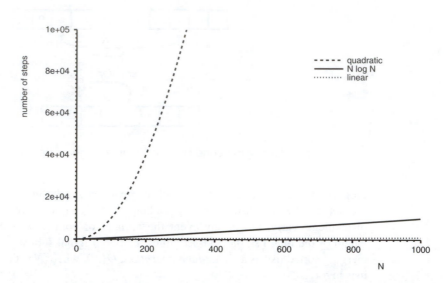

Fig. 5.4 Comparison of N^2, $N \log_2 N$, and N, for N up to 1000. A function that grows at the rate $O(N \log N)$ is much closer to linear than to quadratic.

program. But there's another way to think about merge sort, called *recursive*, or *top-down*, which is elegant and useful. It lets us lean on the compiler to fill in details, allowing us to concentrate on the central principle. The program that results from the recursive coding can use more storage and time than the nonrecursive approach, but the savings we get from having an $O(N\log N)$ algorithm are so dramatic that the inefficiency of the implementation is often irrelevant.

Here's how to think of merge sort from the top down: To sort N numbers, divide the list into two lists, each of size $N/2$; sort each, and then merge. How do we sort each of the half-sized sublists? The same way: divide each in two, sort, and merge. We can express this idea very succinctly with the following code:[†]

```
merge_sort(list)
{
  divide list into two half-size lists;
  merge_sort(first half);
  merge_sort(second half);
  merge the two sorted half-lists;
}
```

[†] I'll use an informal version of the most common procedural languages, closer to C than to Pascal. I hope it's self-explanatory.

The procedure `merge_sort` calls itself — and this top-down, recursive expression of the algorithm will be compiled automatically into the final code. Figure 5.5 shows the recursive idea diagrammatically.

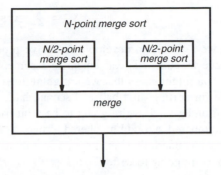

Fig. 5.5 Recursive expression of merge sort: a procedure that calls itself.

We ought to fill in one more detail; otherwise we would encounter disaster when we tried to run such a recursive program. The original procedure call to sort an eight-element list, for example, would generate two calls to sort four-element lists; these would generate four calls to sort two-element lists; these would generate eight calls to sort one-element lists, . . . but how would this process of generating smaller and smaller sublists terminate? We had better notice that sorting a one-element sublist is trivial: we just leave the element as it is and return from the procedure. Thus we should alter the pseudo-code above to include an escape clause when we get down to lists of length one:

```
merge_sort(list)
{
  if (length of list equals one)
    return;

  divide list into two half-size lists;
  merge_sort(first half);
  merge_sort(second half);
  merge the two sorted half-lists;
}
```

This bare skeleton leaves out all the programming details, but captures the main idea. You should be able to flesh it out to produce a working program (Problem 8).

We're now ready to apply divide-and-conquer to the DFT.

6 Decimation-in-time FFT

To apply divide-and-conquer to the DFT we need to split it into two DFTs, each half the size of the original. Here's the DFT, Eq. 2.6, rewritten for convenience:

$$X_k = \sum_{t=0}^{N-1} x_t e^{-jtk2\pi/N} \tag{6.1}$$

Since we're going to develop a recursive procedure based on successive halving of sequence lengths, it is convenient to assume, as in merge sort, that N is a power of two. You might guess that we are going to split this into two parts in the same way we did for sorting: first half and second half. But to derive an FFT algorithm that is easy to understand, we're going to be a bit tricky, and divide the summation into its even-numbered and odd-numbered terms. By indulging in that complication now, we will get a wonderfully simple merge operation.

So we're going to write Eq. 6.1 as

$$X_k = E_k + O_k \tag{6.2}$$

where E_k stands for the sum of the even-numbered points in the DFT, every other term in Eq. 6.1; and O_k for the odd-numbered terms. The even-numbered terms correspond to the index values counting by twos, $t = 0, 2, 4, \ldots, N-2$. To count by ones again, replace t by $2m$, and let $m = 0, 1, 2, \ldots, N/2-1$:

$$E_k = \sum_{m=0}^{N/2-1} x_{2m} e^{-j2mk2\pi/N} \tag{6.3}$$

This is a thinly disguised half-length DFT. Just divide numerator and denominator of the exponent of the complex exponential by two:

$$E_k = \sum_{m=0}^{N/2-1} x_{2m} e^{-jmk2\pi/(N/2)} \tag{6.4}$$

To make it even more obvious, replace $N/2$ by M:

$$E_k = \sum_{m=0}^{M-1} x_{2m} e^{-jmk2\pi/M} \tag{6.5}$$

This is precisely the DFT of the M-point signal $\{x_0, x_2, x_4, \ldots, x_{N-2}\}$, the even-numbered points of the signal x.

Next, rewrite the odd-numbered terms in the DFT Eq. 6.1 by letting $t = 2m + 1$ for the same range of m:

$$O_k = \sum_{m=0}^{N/2-1} x_{2m+1} e^{-j(2m+1)k2\pi/N} \tag{6.6}$$

The manipulation works as before, except we first need to factor out the term corresponding to the extra 1 in the $2m+1$ in the exponent of the complex exponential. We can do this because that factor is independent of the index m:

$$O_k = \sum_{m=0}^{N/2-1} x_{2m+1} e^{-j2mk2\pi/N} e^{-jk2\pi/N} = e^{-jk2\pi/N} \sum_{m=0}^{N/2-1} e^{-j2mk2\pi/N} \tag{6.7}$$

Finally, replace $N/2$ with M, as above:

$$O_k = e^{-jk\pi/M} \sum_{m=0}^{M-1} x_{2m+1} e^{-jmk2\pi/M} \tag{6.8}$$

Once again, this is precisely an M-point DFT, this time of the odd-numbered points of the signal x, namely $\{x_1, x_3, x_5, \ldots, x_{N-1}\}$ — except there is a complex exponential factor in front that depends on k, the frequency index of the transform.

We now have all the ingredients we need for a divide-and-conquer algorithm like merge sort: We have expressed the N-point transform in terms of two $N/2$-point transforms; and once we have the half-sized transforms, we have a simple way to combine the results. The final N-point transform is just the point-by-point sum of Equations 6.5 and 6.8:

$$X_k^{even} + e^{-jk\pi/M} X_k^{odd}, \quad k = 0, 1, 2, \ldots, N-1 \tag{6.9}$$

where X_k^{even} and X_k^{odd} are, respectively, the ($N/2$-point) DFTs of the even- and odd-numbered original input points.

There is one small point that needs some attention, though, before our algorithm is complete. When we compute the half-sized transforms X_k^{even} and X_k^{odd}, we get values at the frequency points corresponding to $k = 0, 1, \ldots, N/2 - 1$. But when we come to combine the two smaller transforms to obtain the N-point transform in Eq. 6.9, we need the results for the full range of k, up to $N-1$. This presents no real problem, however, because the summations X_k^{even} and O_k^{odd} are periodic in k, with period $N/2$. All we need to do to get the values in the range $k = N/2$ to $N-1$ is to replace k by $k - N/2$. The structure of the final FFT algorithm is represented diagrammatically in Fig. 6.1.

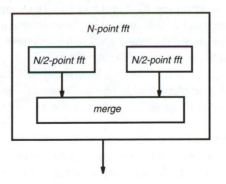

Fig. 6.1 Recursive expression of the FFT; a divide-and-conquer algorithm like merge sort.

Finally, here's the pseudo-code for a recursive FFT function, very much like that for merge sort:

```
fft(signal)
{
 if (length of signal equals one)
  return;

 divide signal into two half-length signals,
  using the even- and odd-numbered points;
 fft(even-numbered signal);
 fft(odd-numbered signal);
 merge the two half-length transforms,
  using Eq. 6.9;
}
```

Notice that we stop the recursion in the divide-and-conquer process when the signal length gets down to one; the length-one transform is trivial: $X_0 = x_0$, which can be checked easily in Eq. 6.1 for $N = 1$.

7 Programming considerations

The recursive algorithm just described can be coded directly in a language like C, and it will produce a fine program, achieving gains in efficiency like those illustrated in Fig. 5.4. It takes $O(N\log N)$ steps, and is nothing to be ashamed of. However, the decimation-in-time FFT is so widely used in practice that its coding has been highly optimized. It is almost always programmed in a nonrecursive manner, merging groups of two, then four, then eight, and so on, without explicit use of recursion. This approach eliminates the many function calls that the compiler supplies to make the recursion happen, and these function calls are relatively time-consuming because of the bookkeeping. (See Problem 10.) The nonrecursive implementation also eliminates a great deal of data movement. Instead of rearranging the data into even and odd parts each time we descend a level in the recursive tree, the data is rearranged once and for all at the beginning of the algorithm.

I programmed the FFT recursively and compared its speed with a standard nonrecursive implementation on transforms of signals of lengths 512 to 4096.[†] The results are listed in Table 7.1, and show that the nonrecursive program is faster by a factor of 3 to 4.

I won't go into great detail about the nonrecursive program, but I do want to mention the main idea involved in getting it to work. Recall that when we first described merge sort in Section 5 we sorted the array *in place* — the elements were rearranged in the same array they originally occupied. It wasn't until we expressed the algorithm recursively that we let the compiler take care of where the elements were stored at the different stages. Now we will go back and see how to implement the FFT in place.

[†] To be more precise, I translated a classic nonrecursive FORTRAN program written by Cooley, Lewis, and Welch into C. See the Notes at the end of this chapter.

signal length, n	recursive time	nonrecursive time	ratio
512	2.1	0.7	3.0
1024	4.7	1.5	3.1
2048	9.9	2.8	3.5
4096	22.3	6.4	4.1

Table 7.1 Comparison between the running times of two FFT algorithms, one recursive and the other nonrecursive. The times shown are reported user CPU times in seconds for 100 repeated FFTs.

In analogy to the in-place version of merge sort, we want to be able to merge two half-size transforms into one full-size transform, and we want the two half-size transforms to sit side by side in the signal array. The merge operation itself simply involves repeatedly forming terms according to Eq. 6.9. There's a difficulty here, however, that isn't present in sorting. Consider the final stage, where we would like to merge two transforms of size $N/2$, one in the first half of the array, and the other in the second half. We need to ensure that the first half contains the transform of the *even*-numbered signals points, and the second half contains the transform of the *odd*-numbered signal points. Otherwise, the merging formula doesn't work. To do this, we should first split up the signal so the even-numbered points sit in the first half of the array and the odd-numbered points in the second half. Therefore, to make the final merge work, we should rearrange the data as shown in the second line of Fig. 7.1, using a 16-point FFT as an example.

The next-to-the-last stage involves merging the first and second quarters of the array, then the third and fourth quarters. But the same problem presents itself, for the same reason: These merges require that we first move the even-numbered points in the first half to the first quarter, and the odd-numbered points in the first half to the second quarter, and similarly in the second half of the array. This rearrangement is shown in the third line of Fig. 7.1.

It should be clear by now that we need to rearrange the data by even/odd splits all the way down to groups of two, to prepare the array for the very first merge stage. The final rearrangement necessary for a 16-point FFT is shown on the bottom line of Fig. 7.1. From this we see that we should rearrange the array by leaving element 0 alone; moving element 8 to position 1; moving element 4 to position 2; and so on. This may seem at first like a ridiculously complicated pattern. How could we possibly figure out how to rearrange 2^{10} points so that all the merges will work with adjacent subarrays? Can you see the pattern relating the first and last rows of Fig. 7.1?

The pattern is actually not that complicated — once you see it, of course. The key to the rearrangement pattern lies in the binary numbers that represent the array indices. For example, consider element 5 (base-10), which has the binary index 0101. The least-significant bit, a 1, means that the element is in an odd-numbered position, so it needs to be put in the second half of the array. The second least significant bit is a 0, which means that among the odd-numbered elements, it occupies an even-numbered position — it is, in fact, in position 2 (counting from 0) among the odd-

```
Start:       0   1   2    3   4    5    6    7   8   9   10   11   12   13   14   15
Stage 1:     0   2   4    6   8   10   12   14 ‖ 1   3    5    7    9   11   13   15
Stage 2:     0   4   8   12 ‖ 2    6   10   14 ‖ 1   5    9   13 ‖  3    7   11   15
Stage 3:     0   8 ‖ 4   12 ‖ 2   10 ‖  6   14 ‖ 1   9 ‖  5   13 ‖  3   11 ‖  7   15
```

Fig. 7.1 Successive stages in rearranging the data for a 16-point FFT. The first line shows the original indices; the second shows the indices after the first even/odd split; and so on. The last line shows the final, bit-reversed reordering.

numbered elements. This means that it will be in the first half of the second half of the array. The next bit is a 1, which means that it will be in the second half of the first half of the second half of the array. Thus, the least-significant bit of the original position determines the most-significant bit of its final position; the second-least-significant bit determines the second-most-significant bit, and so on.

The end result of this argument is that the final position in binary is simply the original position in binary, but with the bits *in reverse order*. Thus, the element in position 5 (base 10), 0101 in binary, winds up in position 10 (base 10), 1010 in binary. An element with an index that in binary is the same reversed, like 9 (base 10), 1001 in binary, stays put, as you can verify in Fig. 7.1.

A nonrecursive FFT program therefore begins with the rearrangement just described, called *bit-reversal*, or *shuffling*. The rest of the program then proceeds by merging adjacent sublists, just as in merge sort, and every sublist will represent the half-size transform of the appropriate even or odd part of corresponding full-size transform. The final nonrecursive program is shown diagrammatically for the eight-point case in Fig. 7.2. The initial bit-reversal rearrangement prepares the data so we can then follow the nonrecursive merge sort program (compare with Fig. 5.2).

Finally, I want to mention one more practical matter in writing an FFT program. The merge steps, in accordance with Eq. 6.9, will consist of a loop of the following form, containing a multiplication by the complex exponential W^k:

```
for (k=0; k<M; k=k+1)
{
(first-half element) + W^k *(second-half element);
}
```

This loop is, in fact, the inside loop, and contains all the arithmetic in the entire algorithm. It is therefore important that we pay close attention to its efficiency, and notice that it would be silly to recompute the power of W from scratch every time we needed it. It is much more efficient to generate each successive power by multiplying the previous one by W, as in the following:

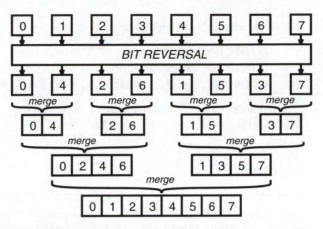

Fig. 7.2 Outline of the nonrecursive eight-point FFT program. The merge steps follow Eq. 6.9.

```
U = 1;
for (k=0; k<M; k=k+1)
{
(first-half element) + U*(second-half element);
U = U*W;
}
```

8 The inverse DFT

We'll want to compute the inverse DFT as well as forward DFT, but we don't need a separate program to do it. The two are almost identical, as you can see by comparing Eqs. 2.5 and 2.6. The inverse transform has an extra factor of $1/N$ — just a constant scale factor — and the exponent of the complex exponential is the negative of the one in the forward transform.

We can see how to use the forward DFT to compute the inverse DFT by a simple manipulation of the inverse transform, Eq. 2.5:

$$x_t = \frac{1}{N} \sum_{k=0}^{N-1} X_k e^{jtk2\pi/N} \tag{8.1}$$

Take the complex conjugate of both sides of this equation and multiply by N:

$$N x_t^* = \sum_{t=0}^{N-1} X_k^* e^{-jtk2\pi/N} \tag{8.2}$$

The right-hand side is just the forward DFT of X_k^*. This tells us that the forward DFT of the conjugate of the transform gets us back to something quite close to the original signal; namely, Nx_t^*. To get back to x_t, all we need to do is divide by N and take the complex conjugate.

To summarize, to calculate the inverse DFT a signal:

(a) take its conjugate;
(b) take its forward transform;
(c) take the conjugate of the result and divide by N.

The extra operations take a number of steps proportional to N. I ask you to derive another efficient method in Problem 15.

9 A serious problem

You might think that we are now in a great position to use the FFT. Let's say we want to determine the frequency content of a signal. We take a sample of, say, 1024 (a power of 2) consecutive points, apply our $O(N\log N)$ FFT algorithm, and look at the magnitude of the DFT it computes as a function of frequency. It turns out that there is more to it than that, and the difficulties are interesting and instructive. They arise because we are taking a slice of a signal that almost always extends before and after our sample. To do things right we need to study the frequency transform of signals that extend to the infinite past and infinite future. This we'll do in the next chapter, after which we'll return to the practical application of the FFT, armed with a better understanding of what the frequency content determined by the DFT really means.

Before we delve into the case of infinite-extent signals, however, I want to illustrate the problems that come up when we take the DFT of a piece of a sinusoid. To make things simple, suppose we take a sample of $N = 1024$ consecutive points of a particular basis element, the complex phasor with frequency ($133/N$) times the sampling frequency:

$$x_t = e^{jt(133)2\pi/N} \tag{9.1}$$

We know from our previous work that its DFT is zero everywhere except at point 133, where it is $N = 1024$ (recall Eq. 2.4 and see Problem 1). That's the theory — but when we compute the DFT with the FFT, there will be small computational errors introduced by roundoff, as we'll see presently.

The output of the FFT consists of 1024 points, the frequency content at the sampling frequency times $i/1024$, $i = 0, 1, \ldots, 1023$. It's therefore convenient to use $i/1024$ as the abscissa, labeled ''fractions of the sampling frequency,'' as shown in Fig. 9.1. That figure shows the DFT as computed with the FFT algorithm, plotted as magnitude in dB. Instead of true zero at the points excluding the true phasor frequency, we get numbers below $10^{-11} = -220$ dB. But this is what we would expect; floating-point arithmetic isn't perfect, and, after all, we're getting a range of maximum to minimum amplitude of about 280 dB, or 14 decimal places.

Suppose, though, that the phasor we analyze happens to be at a frequency 133.5/1024 times the sampling frequency. In other words, suppose its frequency falls in the ''crack'' between two DFT points. The result is shown in Fig. 9.2, with a drastically different ordinate scale. The DFT now does a very poor job of resolving the

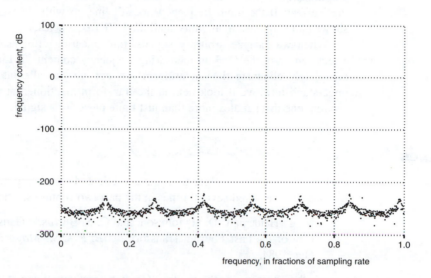

Fig. 9.1 The 1024-point FFT of a basis element, the phasor with frequency precisely 133/1024 = 0.12988 times the sampling frequency. Don't forget to notice the lonely point 133 with ordinate 1024, or 60.2 dB.

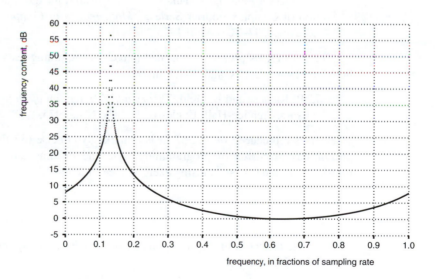

Fig. 9.2 The 1024-point FFT of a phasor with frequency 133.5/1024 = 0.13037 times the sampling frequency.

frequency of this particular input signal. In fact, the range of maximum to minimum amplitude is 58 dB, or about three decimal places. You can think of this as a ''smearing'' effect: If the input frequency doesn't line up with a basis element, its energy shows up in significant amounts at all the DFT frequencies.

When we analyze arbitrary signals, this smearing effect seriously degrades the effectiveness of the DFT in measuring frequency content. It turns out that we can't completely eliminate the smearing effect, but we can alleviate it significantly. To understand how, we'll look next at the transform mentioned at the beginning of this section, one that handles more than just finite pieces of a signal.

Notes

Fast Fourier Transform algorithms have a long and interesting history. The basic idea has been rediscovered several times in the past two centuries. The following paper:

> M. T. Heideman, D. H. Johnson, and C. S. Burrus, ''Gauss and the History of the Fast Fourier Transform,'' *IEEE ASSP Magazine*, pp. 14–21, Oct. 1984.

traces that history back to the German mathematician Carl Friedrich Gauss (1775–1855) who described it in a Latin treatise, *Theoria Interpolationis Methodo Nova Tractata,* most likely written in 1805, but not published until 1866. The explosion in its use for signal processing was detonated by the following paper, which describes a version applicable to any sequences with nonprime lengths:

> J. W. Cooley and J. W. Tukey, ''An Algorithm for the Machine Calculation of Complex Fourier Series,'' *Mathematics of Computation*, vol. 19, no. 2, pp. 297–301, April 1965.

The FORTRAN program translated into C for the timing tests in Section 7 is attributed to Cooley, Lewis, and Welch in

> L. R. Rabiner and B. Gold, *Theory and Application of Digital Signal Processing,* Prentice-Hall, Englewood Cliffs, N.J., 1975,

where it is reproduced on p. 367. It is remarkably concise (32 lines) and violates all present-day standards of programming style and documentation. Deciphering and translating it into C is an amusing exercise (see Problem 12).

Problems

1. Complete the check of the orthogonality of the DFT basis by showing the inner product in Eq. 2.4 is zero if $m \neq n$.

2. Check that taking the inverse DFT of the forward DFT gets back to the original signal by substituting Eq. 2.6 in Eq. 2.5. This shows that using the factor $1/N$ in the inverse transform is consistent with using no factor in the forward transform.

3. Show that the matrix of the DFT transformation, F in Eq. 2.7, is nonsingular. What is its determinant? What is its inverse?

4. Assume you are given a continuous-time signal $x(t)$ that is periodic with period T. That is, $x(t) = x(t+iT)$ for all integers i. Assume further that you sample every T_s sec, and that the signal $x(t)$ has no frequency components beyond the Nyquist frequency, $1/(2T_s)$ Hz. (When that is true, we say $x(t)$ is *bandlimited*.) Finally, assume that the sampling interval T_s is a divisor of the repetition period T, and let $N = T/T_s$, the number of samples in a basic period.

(a) Derive the DFT and inverse DFT of the digital signal consisting of the first N samples in terms of the Fourier series of $x(t)$.

(b) Explain how the result is changed when the original signal is not bandlimited.

5. In the analysis of the running time of merge sort, I point out that the number of times we have to double the sizes of lists from one to get past N is about $\log_2 N$. Give the precise number of stages in the algorithm.

6. Write a program in your favorite language that computes the DFT using the decimation-in-time FFT algorithm described in Section 6.

7. Write another program that computes the DFT in $O(N^2)$ time, using a naive implementation of the nested loops in Eq. 2.6. Then compare the fast and slow algorithms on data lengths ranging from $N = 2$ to $N = 2048$. For what values of N is the FFT faster?

8. Write a program for merge sort using the recursive outline in Section 5. Run timing tests to verify that the running time becomes proportional to $N\log N$ for large N.

9. Write a program for merge sort that is expressed nonrecursively instead of recursively. How does its running time compare with the recursive implementation in the previous problem?

10. How many function calls are required by a recursive implementation of the decimation-in-time FFT? Express your answer in terms of the signal length N.

11. Write and test an explicitly recursive FFT in C or Pascal.

12. Read, disentangle, and translate into C the Cooley-Lewis-Welch FORTRAN FFT referred to in the Notes. Compare its running times with those of the recursive program in Problem 11. Do you get the same results I did?

13. FORTRAN has a complex data type, so the complex multiplication at the end of Section 7 looks like this in the Cooley-Lewis-Welch program mentioned in the previous problem:

```
U = U*W
```

When I translated this to C, I used `Ur` and `Ui` to represent the real and imaginary parts of `U`, respectively, and similarly for `W`. I then translated the line above into C as

```
Ur = Ur*Wr - Ui*Wi;
Ui = Ur*Wi + Ui*Wr;
```

Find the bug.

14. If we reuse an FFT program many times, which we commonly do, we may want to precompute and store all the sine and cosine values required by the algorithm. This trades space for time. How many values do we really need to store for an N-point FFT, taking into account the symmetries of the sinusoids?

15. The DFT of a vector is represented by multiplication by the matrix F in Eq. 2.7. Consider the effect of taking the DFT twice in succession. This is represented by multiplication by the matrix $R = F^2$.

(a) Find an explicit formula for the ijth element of R.

(b) Show that $R^2 = I$, the identity matrix. What does this imply about R^{-1}? About F?

(c) Find an expression for F^{-1} in terms of F and R, thus suggesting an efficient way to compute the inverse DFT using the forward FFT algorithm — an alternative to the method described at the end of Section 8.

16. The sample FFT plot of a basis element in Fig. 9.1 was obtained using double-precision arithmetic. If you have written or can get an FFT program, compare the results on the same example using single- and double-precision arithmetic. Also try transforming different basis elements, including the ones at points 128 and 256. The results may surprise you.

17. Take a look at the FFT in Fig. 9.2 of a phasor at a frequency lying exactly halfway between two DFT points. Notice that its minimum value is almost precisely 0 dB, or one. Explain this.

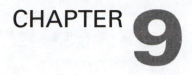

CHAPTER 9

The *z*-transform and Convolution

1 Domains

We've already seen two frequency transforms, both of them for finite-extent (or periodic) signals — Fourier series, for continuous-time signals, and the DFT, for discrete-time signals. In the case of Fourier series, illustrated diagrammatically in Fig. 1.1, the time domain is continuous and finite, and as we discussed in the previous chapter, a finite time domain can be thought of as a circle. On the other hand, the Fourier series spectrum consists of an infinite sequence of complex numbers, indicated in the figure by a dotted straight line.

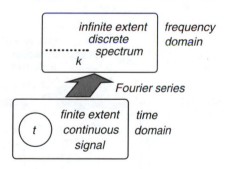

Fig. 1.1 Fourier series as a frequency transform.

In the case of the Discrete Fourier Transform, both the time and frequency domains are finite and discrete. This is illustrated in Fig. 1.2, where both domains are represented by dotted circles.

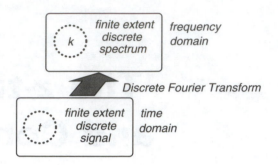

Fig. 1.2 The DFT.

If you think about it for a second, there are just two more possible time domains to consider: the cases where the time axis is infinite in extent and either discrete or continuous. The first of these remaining cases, the z-transform, is the one we'll consider next and is illustrated in Fig. 1.3. This is the important situation where the signal is digital and extends indefinitely in one or both directions. It turns out that the frequency domain of the z-transform is finite and continuous. In fact, the mathematics is really the same as that for the Fourier series, except the time and frequency domains are interchanged. You can think of the z-transform as the inverse of the Fourier series operation — it starts with an infinite sequence and yields a periodic function of a continuous variable, in this case the frequency content for frequencies up to the Nyquist frequency. We could make use of the work we did for the Fourier series to derive the z-transform, but it's so important and so easy that we'll do it from scratch, in the next section.

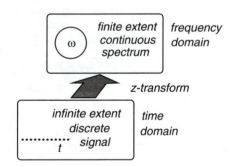

Fig. 1.3 The z-transform.

The final case, the Fourier transform, is illustrated in Fig. 1.4. We already alluded to this situation at the end of Chapter 3 — the signal is not sampled and extends indefinitely in time. In this case the frequencies required to represent signals must also be continuous and extend indefinitely. Both the time and frequency domains are continuous lines of infinite extent. We'll return to the Fourier transform after we discuss the *z*-transform.

The continuous-time cases, the Fourier transform and Fourier series, are the classical, nineteenth-century frequency transforms — they're tools developed by physicists. Sampling is a mid-twentieth-century idea, and the DFT and *z*-transforms are relatively recent inventions. But, as you can see from the way the pieces fit together, the four frequency transforms outlined here are really just different incarnations of the same basic idea: Signals can be decomposed into sums of phasors. The intuition you develop in any of the domains will almost always apply in the others.

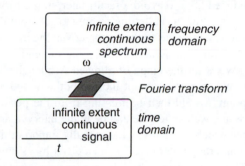

Fig. 1.4 The Fourier transform.

2 The *z*-transform

We're ready to look at the *z*-transform, used for discrete-time signals that are defined for a time axis of infinite extent. We'll rely on the same geometric picture we used before: A signal will be thought of as a point in a space with coordinates that correspond to its different frequency components.

Let's start with the basis, the coordinate axes. These must be discrete-time phasors, signals of the form

$$e^{jk\omega} \tag{2.1}$$

where k is the (discrete) time variable, and ω is the frequency variable in radians per sample.[†] Since we now want to represent signals defined for the infinite time axis, we allow the integer k to vary from $-\infty$ to $+\infty$.

We now come to a key point. The signals represented by the *z*-transform are sampled, so the basis needs to contain only frequencies up to the Nyquist — but no higher. Therefore, we restrict ω to lie between $-\pi$ and $+\pi$ radians per sample. There are a lot of basis elements — one for each real frequency ω between $-\pi$ and π. The representation of the signal f_k is therefore a summation of the phasors in the basis over this

---

[†] At the risk of being repetitious, let me remind you that we could use ωT_s in the exponent, where T_s is the sampling interval, in which case the units of ω would be radians per sec. It's easier to use normalized frequency in radians per sample; the Nyquist frequency is π radians per sample.

range of frequencies, weighted by the amount of the signal at each particular frequency. This summation over a continuum like this is represented by the integral

$$f_k = \frac{1}{2\pi} \int_{-\pi}^{\pi} F(\omega) e^{jk\omega} \, d\omega \qquad (2.2)$$

The function $F(\omega)$ is the *frequency content*, or, loosely speaking, the *spectrum* of the signal f_k. In general $F(\omega)$ is complex; it tells us not only the magnitude but also the phase angle of the phasor at frequency ω present in the signal. The factor $1/(2\pi)$ is introduced here to avoid a factor later, in the forward transform — Eq. 2.2 is actually an inverse transform, telling us how to get from the spectrum back to the signal. This is exactly analogous to the factor $1/N$ in the inverse DFT, Eq. 2.5 of Chapter 8.

One small point here, concerning the end points of the band of allowed frequencies. We want the range of allowed frequencies to repeat periodically, every 2π, forever. We therefore cannot include both $-\pi$ and π in the allowed band, because these endpoints would then be missing from the periodically shifted versions. To resolve this issue, we can let one end be in the interval, and the other end outside — it doesn't matter which. Actually, what is really important is that no two frequencies in the fundamental range between $-\pi$ and π differ by as much as 2π.

The next step is to find the appropriate inner product, which in our work is always a summation of products over the time variable. Since the time variable is discrete and infinite in extent, we have little choice; the inner product between two discrete signals f_k and g_k must be

$$\langle f_k, g_k \rangle = \sum_{k=-\infty}^{\infty} f_k g_k^* \qquad (2.3)$$

Once more, we use the complex conjugate of the second signal so the inner product of a signal with itself is real, the sum of the squares of absolute values. Notice that in general we allow signals to extend to negative time, although we will often consider signals that are zero for $k < 0$, so-called *one-sided* signals.

We've followed the same program twice before, so it should be familiar. As mentioned above, the signal representation in terms of frequency content, Eq. 2.2, is the inverse transform. It goes from the frequency domain, the frequency content $F(\omega)$, to the time-domain signal f_k. The forward transform should express the coordinate values $F(\omega)$ in terms of the signal, using the inner product. To find the content of the signal f_k at frequency ω, take the inner product of f_k with the basis signal at that frequency:

$$F(\omega) = \langle f_k, e^{jk\omega} \rangle = \sum_{k=-\infty}^{\infty} f_k e^{-jk\omega} \qquad (2.4)$$

The frequency ω will range from $-\pi$ to π, so the complex exponential $e^{j\omega}$ ranges exactly once over the unit circle.

We now come to the definition of the z-transform. Although we've been considering values of z on the unit circle ($z = e^{j\omega}$), we're going to get a lot of mileage out of considering z to be a full-fledged complex variable. We'll therefore make the substitution

$$z = e^{j\omega} \tag{2.5}$$

in the forward transform in Eq. 2.4, which then becomes the following simple-looking power series:

$$\mathcal{F}(z) = \sum_{k=-\infty}^{\infty} f_k z^{-k} \tag{2.6}$$

This is the *z-transform*. We'll see shortly that this z has precisely the same meaning as the z introduced in Chapter 4 to analyze the frequency response of digital filters.

Notice that I've gone out of my way to use different symbols for the functions $F(\omega)$ and $\mathcal{F}(z)$. Actually, they're related by

$$F(\omega) = \mathcal{F}(e^{j\omega}) \tag{2.7}$$

The z-transform of a signal evaluated on the unit circle tells us its frequency content. We used the same notation in Chapters 4 and 5 for digital filter transfer functions, and, as we'll see soon, the two situations are very closely related. The z-transform on the unit circle, Eq. 2.4, is sometimes referred to as the *Discrete-Time Fourier Transform* (*DTFT*), but we won't bother using the extra terminology.

Before we move ahead, we should fill in a missing piece of the mathematics: the verification that the basis is orthogonal. The next section deals with that issue. You can skip it without losing continuity, but the way the δ functions from Chapter 7 come into play here is elegant, and a good example of time-frequency symmetry.

3 Orthogonality

We skipped checking the orthogonality of the basis we used in the z-transform, so we're going to check it now. The geometric intuition behind orthogonality of basis elements is that the two vectors representing two different elements are at right angles to each other. Remember that the basis elements are signals indexed by their frequency ω — there is one basis element for each frequency in the range $-\pi$ to π. So let's consider the inner product between two basis elements that correspond to frequencies ω_1 and ω_2. Substituting these from Eq. 2.1 into the inner product formula Eq. 2.3 gives us

$$\langle e^{jk\omega_1}, e^{jk\omega_2} \rangle = \sum_{k=-\infty}^{\infty} e^{jk(\omega_1 - \omega_2)} \tag{3.1}$$

Ordinarily, the right-hand side of this equation would stop us in our tracks. There seem to be real problems here. When $\omega_1 = \omega_2$, for example, the right-hand side of Eq. 3.1 is infinite, the sum of an infinite number of ones. When $\omega_1 \neq \omega_2$, we have an equally unsatisfactory situation: The terms are complex, all of unit magnitude (by Euler's formula), and each term is equal to the one before, except rotated by

$(\omega_1 - \omega_2)$. It would be hard to make sense of this, were it not for the fact that we've already seen this kind of expression in Chapter 7, when we were looking at buzz and the Fourier series for a square wave and its derivative. With a little work we can convert that result into just what we need now.

Equation 6.2 of Chapter 7 tells us that

$$\sum_{k=-\infty}^{\infty} e^{jk2\pi t/T} = \sum_{k=-\infty}^{\infty} T\delta(t - kT) \qquad (3.2)$$

where we have replaced the ω_0 in Chapter 7 by $2\pi/T$. As a function of t, this is a train of Dirac δ functions, each with area T, spaced T sec apart. We want the left-hand side of Eq. 3.2 to be exactly like the right-hand side of Eq. 3.1, so we next choose $T = 2\pi$, to match the constants in the complex exponent. (Since Eq. 3.2 is true for all T, we're free to choose T to be anything we want.) Furthermore, the sum in Eq. 3.2 is considered a function of t, but we want to regard Eq. 3.1 as a function of ω. Therefore, replace t in Eq. 3.2 by ω, yielding[†]

$$\sum_{k=-\infty}^{\infty} e^{jk\omega} = 2\pi \sum_{k=-\infty}^{\infty} \delta(\omega - k2\pi) \qquad (3.3)$$

This is, actually, just what we're looking for, except that the ω on the left-hand side will be set equal to $(\omega_1 - \omega_2)$ in order to match Eq. 3.1. However, because ω_1 and ω_2 are both restricted to the range between $-\pi$ and π we can put the right-hand side in a simpler form.

The δ function $\delta(\omega)$ has its spike of infinite height when $\omega = 0$. The spikes on the right-hand side of Eq. 3.3 therefore occur when ω_1 and ω_2 differ by an integer multiple of 2π. Now recall our observation in Section 2 that we should exclude one of the endpoints of the interval between $-\pi$ and π — which implies that ω_1 and ω_2 always differ by less than 2π. Therefore, only one spike occurs on the right-hand side of Eq. 3.3, when $\omega_1 = \omega_2$ and $k = 0$.

Combining Eq. 3.1, the original inner-product calculation, with the simplified Eq. 3.3 yields:

$$\langle e^{jk\omega_1}, e^{jk\omega_2} \rangle = 2\pi \, \delta(\omega_1 - \omega_2) \qquad (3.4)$$

This is the orthogonality result we need.

We can now make sense of the problems we observed above. It is true that the inner product between two basis signals with the same frequency is infinite. It also turns out that the inner product when the frequencies differ is zero. What we get, in fact, as a function of the difference in frequencies, is a spike with an area of 2π. And this is exactly what we need to derive the forward transform, Eq. 2.4, from the inverse transform, Eq. 2.2, our signal representation and starting point.

To see this, let's start with what we claim is the frequency content of our original signal — its inner product with a basis signal at frequency ω':

$$\langle f_k, e^{jk\omega'} \rangle \qquad (3.5)$$

[†] This is neither the first nor the last time we're going to interchange time and frequency variables. Time-frequency symmetry will become one of our Leitmotifs.

Next, recall the inverse transform, Eq. 2.2, for f_k:

$$f_k = \frac{1}{2\pi} \int_{-\pi}^{\pi} F(\omega)\, e^{jk\omega}\, d\omega \tag{3.6}$$

and use this in Eq. 3.5:

$$\langle f_k,\, e^{jk\omega'} \rangle = \frac{1}{2\pi} \int_{-\pi}^{\pi} F(\omega) \langle e^{jk\omega},\, e^{jk\omega'} \rangle\, d\omega \tag{3.7}$$

Notice that, with our usual mathematical abandon, we've slid the inner product inside the integral. In other words, we've interchanged the order of the integration and the summation represented by the inner product. A felicitous collapse now ensues. The inner product becomes the δ function we worked to get in Eq. 3.4, which then punches out the integrand at the frequency $\omega = \omega'$, at the same time canceling the $1/(2\pi)$ factor in front of the integral. The net result is simply

$$F(\omega') = \langle f_k,\, e^{jk\omega'} \rangle \tag{3.8}$$

which is precisely Eq. 2.4, the forward transform we wrote down from geometric intuition.

I hope you didn't mind the excursion in this section. It shows that the pieces we've been using fit together the way they should, and it gave us a bit more practice using δ functions.

4 *z*-transform of the impulse and step

It's time to get down to specifics and look at some important examples of *z*-transforms. This will sharpen your intuition and give you a drawerful of useful standard parts to use later. Our "simplest" signal up to now has been the phasor; but now, for a change, let's start with another very simple signal, called the *unit sample*, or *unit impulse* digital signal, which is illustrated in Fig. 4.1. In fact, nothing could be simpler — it's one at the zeroth sample, and zero everywhere else:

$$\delta_k = \begin{cases} 1 & \text{if } k = 0 \\ 0 & \text{if } k \neq 0 \end{cases} \tag{4.1}$$

The *z*-transform of this signal is the power series Eq. 2.6, and only the zeroth term contributes anything. But the zeroth term is multiplied by $z^0 = 1$, so the *z*-transform of the unit sample signal is just 1. That is, all frequencies are present in equal amounts. We should now try to make sense out of this possibly puzzling fact.

First, let's go back to our other frequency representations to see if their results jibe with this one. The periodic continuous-time signal analogous to a single unit sample is a train of repeating δ functions. But we've already seen that in Chapter 7. What's more, we used it just recently, in Eq. 3.2, which we rewrite here as

$$\sum_{k=-\infty}^{\infty} \delta(t - kT) = \frac{1}{T} \sum_{k=-\infty}^{\infty} e^{jk2\pi t/T} \tag{4.2}$$

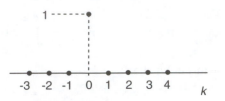

Fig. 4.1 The unit sample digital signal.

All the Fourier coefficients are the same, $1/T$. Check!

To find the analogy in the DFT case, we should look at the signal

$$f_k = \begin{cases} 1 & \text{if } k = 0 \\ 0 & \text{if } k = 1,\, 2,\, 3,\, \ldots,\, N-1 \end{cases} \tag{4.3}$$

The summation defining the DFT, Eq. 2.6 in Chapter 8, becomes simply

$$F_k = 1 \,, \quad \text{for } k = 0,\, 1,\, 2,\, 3,\, \ldots,\, N-1 \tag{4.4}$$

Again, check!

You can get some intuition for these transforms by going back to our work on Fourier series in Chapter 7. We observed then that the more abrupt the changes in a time function, the more high frequencies are required to represent it. Thus, the isolated pulse, being an extremely abrupt change, requires all frequencies in equal amounts.

Recall from Section 3 of Chapter 7 that representing a square wave requires frequencies in amounts inversely proportional to their frequency. In other words, the nth harmonic is represented with the factor $1/n$ (see Eq. 3.5 in Chapter 7). Now let's try to verify that result for the z-transform. Figure 4.2 shows the discrete-time, infinite-extent unit step signal,

$$u_k = \begin{cases} 1 & \text{if } k \geq 0 \\ 0 & \text{if } k < 0 \end{cases} \tag{4.5}$$

The z-transform is the series

$$\mathcal{U}(z) = \sum_{k=0}^{\infty} z^{-k} \tag{4.6}$$

This is a geometric series, one we've already met in Chapter 7 (see Problem 7 of that chapter). The sum is

$$\mathcal{U}(z) = \frac{1}{1 - z^{-1}} \tag{4.7}$$

The frequency content of the unit step signal is the magnitude of its z-transform on the unit circle in the z-plane, shown plotted in Fig. 4.3. The plot shows that the frequency content peaks at zero frequency (DC), and decreases as the frequency increases, which is consistent with our expectation that it vary as the inverse of

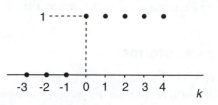

Fig. 4.2 The unit step digital signal.

frequency. In fact, there appears to be an infinite amount of DC present, which makes sense for the following reason. Zero frequency corresponds to the value $z = e^0 = 1$. Setting $z = 1$ in the definition of the *z*-transform tells us that

$$\mathcal{U}(1) = \sum_{k=-\infty}^{\infty} u_k = \sum_{k=0}^{\infty} 1 = \infty \tag{4.8}$$

This is a divergent series, and the zero-frequency content is infinite. Put more simply, the function $\mathcal{U}(z)$ has a *pole* at $z = 1$.

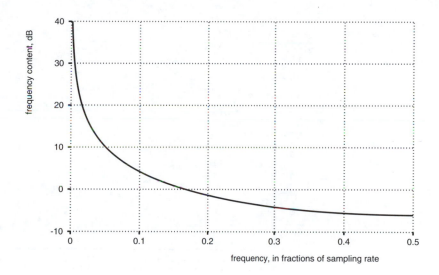

Fig. 4.3 Frequency content of the unit step digital signal.

To check the shape of the frequency content curve more closely, set $z = e^{j\omega}$ in $\mathcal{U}(z)$, yielding

$$|U(\omega)|^2 = \frac{1}{(1 - \cos\omega)^2 + \sin^2\omega} = \frac{\frac{1}{2}}{1 - \cos\omega} \tag{4.9}$$

When ω is small, $\cos\omega \approx 1 - \omega^2/2$, so for small ω

$$|U(\omega)| \approx 1/\omega \tag{4.10}$$

Notice that the inverse-frequency shape checks the Fourier series result for small ω, but not for frequencies near Nyquist. I'll ask you to think about that in Problem 1.

5 A few more *z*-transforms

It's very easy to derive additional useful *z*-transforms from the unit step transform, which you'll recall from the previous section is the geometric series

$$\mathcal{U}(z) = \sum_{k=0}^{\infty} z^{-k} = \frac{1}{1 - z^{-1}} \tag{5.1}$$

Suppose we consider the same signal, but with each sample value weighted by an exponential factor R^k:

$$f_k = \begin{cases} R^k & \text{if } k \geq 0 \\ 0 & \text{if } k < 0 \end{cases} \tag{5.2}$$

This has an exponentially decaying shape, as shown in Fig. 5.1. The closer R is to one, the slower the decay.

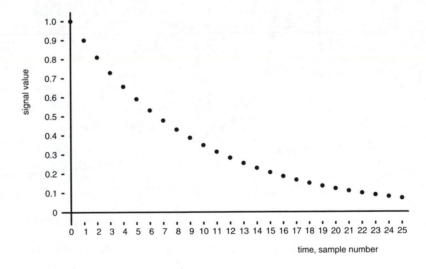

Fig. 5.1 Exponentially damped step signal, $R = 0.9$.

The *z*-transform is

$$\mathcal{F}(z) = \sum_{k=0}^{\infty} R^k z^{-k} = \sum_{k=0}^{\infty} (R^{-1}z)^{-k} \tag{5.3}$$

This is a geometric series, but with z replaced by $R^{-1}z$. Therefore its z-transform is just Eq. 5.1 with that change of variable:

$$\mathcal{F}(z) = \frac{1}{1 - Rz^{-1}} \tag{5.4}$$

Note that this moves the pole from $z = 1$ to $z = R$.

There is nothing to prevent us from using an exponential weighting factor that is complex. Consider the signal

$$f_k = \begin{cases} R^k e^{jk\theta} & \text{if } k \geq 0 \\ 0 & \text{if } k < 0 \end{cases} \tag{5.5}$$

where we are now free to choose two parameters, R and θ, which, as you might expect, determine the radius and angle of a pole in the z-plane. The same procedure as above leads to the z-transform

$$\mathcal{F}(z) = \frac{1}{1 - Re^{j\theta}z^{-1}} \tag{5.6}$$

And there's the pole: at $z = Re^{j\theta}$.

The complex exponential signal in Eq. 5.5 can be thought of as the sum of two signals by taking the real and imaginary parts:

$$\mathcal{Real}\,\{f_k\} = R^k \cos(k\theta), \qquad \text{for } k \geq 0 \tag{5.7}$$

and

$$\mathcal{Imag}\,\{f_k\} = R^k \sin(k\theta), \qquad \text{for } k \geq 0 \tag{5.8}$$

Figure 5.2 shows an example of the damped cosine wave.

We can get the z-transforms of each of these by breaking down the transform in Eq. 5.6 the same way. To do this, multiply the numerator and denominator of Eq. 5.6 by the denominator with j replaced by $-j$.[†] The result is

$$\mathcal{F}(z) = \frac{1 - (R\cos\theta)z^{-1}}{1 - (2R\cos\theta)z^{-1} + R^2 z^{-2}} + j\,\frac{(R\sin\theta)z^{-1}}{1 - (2R\cos\theta)z^{-1} + R^2 z^{-2}} \tag{5.9}$$

The first and second parts of this equation are the z-transforms of the damped cosine signal and damped sine signal, Eqs. 5.7 and 5.8, respectively. The z-transforms we derived in this section are collected in Table 5.1.

[†] Don't be confused by the fact that z is a complex variable. In this context we can treat it simply as a place marker in the power series that defines the z-transform. We're just breaking the power series into two parts, one with a j in front.

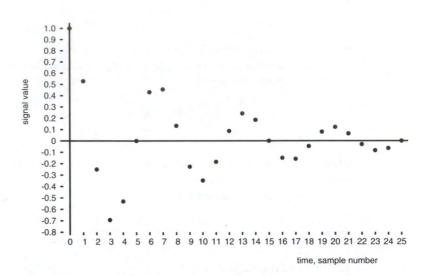

Fig. 5.2 Exponentially damped cosine signal, $R = 0.9$, $\theta = 0.3\pi$.

Digital signal		z-transform
unit impulse	δ_k	1
unit step	u_k	$\dfrac{1}{(1 - z^{-1})}$
damped step	$R^k u_k$	$\dfrac{1}{(1 - Rz^{-1})}$
damped cosine-wave	$R^k \cos(k\theta)$	$\dfrac{1 - (R\cos\theta)\, z^{-1}}{1 - (2R\cos\theta)\, z^{-1} + R^2 z^{-2}}$
damped sine-wave	$R^k \sin(k\theta)$	$\dfrac{(R\sin\theta)\, z^{-1}}{1 - (2R\cos\theta)\, z^{-1} + R^2 z^{-2}}$

Table 5.1 Some useful z-transforms. The last two entries have poles at $z = e^{\pm j\theta}$.

6 *z*-transforms and transfer functions

I hope you're not disturbed at this point by a feeling of *déjà vu*. Of course we've used *z*-transforms before, without calling them that, when we looked at the transfer functions of digital filters. We've also seen the transforms in Table 5.1 before as transfer functions of feedback digital filters. The connection should be clear by now:

- The value of the *z*-transform on the unit circle at the frequency ω represents the amount of the signal present at that frequency;

- The value of the transfer function at the same point represents the effect of the filter on that frequency.

Thus, if the frequency content of a signal is $X(\omega)$, and the frequency response of a digital filter is $H(\omega)$, the output of the filter should have the frequency content $Y(\omega) = H(\omega) \cdot X(\omega)$. It does not take a great leap of imagination to guess that the *z*-transform of the output signal of the digital filter is

$$\mathcal{Y}(z) = \mathcal{H}(z) \cdot X(z) \tag{6.1}$$

This is so important I've drawn a picture of it, Fig. 6.1, even though it's exceedingly simple.

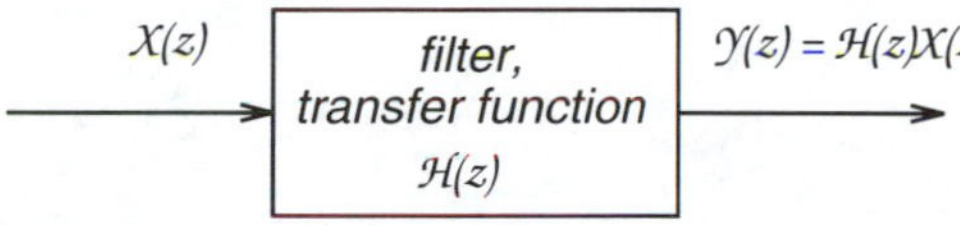

Fig. 6.1 The output *z*-transform is the input *z*-transform multiplied by the transfer function.

To see why this is true, recall how we derived the transfer function in Section 5 of Chapter 4. We assumed there that the input signal was a phasor with fixed frequency ω, and interpreted z^{-1} as a delay operator. Exactly the same technique works for general input signals: multiplying a signal's *z*-transform by z^{-1} is equivalent to delaying the signal by one sample period. To see this, just observe what happens when we multiply the *z*-transform $X(z)$ of x_k by z^{-1}:

$$z^{-1} X(z) = z^{-1} \sum_{m=-\infty}^{\infty} x_m z^{-m} = \sum_{m=-\infty}^{\infty} x_m z^{-(m+1)} = \sum_{k=-\infty}^{\infty} x_{k-1} z^{-k} \tag{6.2}$$

where we got the last equality by replacing $(m+1)$ by k. This means exactly what we want it to mean: the *z*-transform of the delayed signal, x_{k-1}, is $z^{-1} X(z)$. The derivations of the transfer functions of both feedforward and feedback filters go through just as they did in Chapters 5 and 6, except now the results are true for arbitrary input signals, not just phasors.

This is a very fundamental fact. Filtering in the time domain is represented by multiplication by the transfer function in the frequency domain. This is true not only for the frequency content of a signal, determined by the z-transform on the unit circle, but for its entire transform as a function of z.

We now have a new interpretation of the transfer function $\mathcal{H}(z)$ of a digital filter. If the input signal x is a unit impulse, then $X(z) = 1$, and therefore the z-transform of the output signal is $\mathcal{Y}(z) = \mathcal{H}(z)$. Thus *the transfer function can be viewed as the z-transform of the filter's impulse response.*

7 Convolution

The output of the filtering process is determined by two discrete signals: the input signal and the impulse response of the filter. Let's look at exactly how this takes place. The output signal y_t at time t is actually the result of the filter being hit with input samples at all times. For example, the contribution at time t due to the input sample x_0 is

$$x_0 h_t \tag{7.1}$$

The contribution at time t due to x_1 is

$$x_1 h_{t-1} \tag{7.2}$$

The typical contribution due to x_k is $x_k h_{t-k}$. Adding all these up, we see that the total output signal at time t is

$$y_t = \sum_{k=-\infty}^{\infty} x_k h_{t-k} \tag{7.3}$$

You can think of this in fairly abstract terms. It is a recipe for taking two digital signals, x and h, and producing a third. When viewed this way, as a binary operator, the operation is called *convolution*, and is represented by the special symbol $*$, as in:

$$y = x * h \tag{7.4}$$

The limits on the convolution sum, Eq. 7.3, go from $-\infty$ to ∞, but that's just a convenient way to write the convolution mathematically without worrying about special cases. In practical situations the actual number of terms *computed* in the sum is finite. First, the input signal must start at some definite time, say m. Second, even if we have the future input signal available for the indefinite future, the filter can respond to values only some fixed number of samples, say p, beyond the present time. The convolution sum then becomes

$$y_t = \sum_{k=m}^{p} x_k h_{t-k} \tag{7.5}$$

In many cases it's convenient to say that the input signal begins at $k = 0$. And in many cases the filter does not respond to any input values beyond the present time, so $h_{t-k} = 0$ when $t - k < 0$. The convolution is then simply

$$y_t = \sum_{k=0}^{t} x_k h_{t-k} \tag{7.6}$$

which is the way it's often written. Note that it is entirely possible that a digital filter respond to values of input in the "future" — we may just happen to know the entire input before we start filtering. Even simpler is the situation where we can tolerate a delay, waiting until time $t + 10$, say, to produce the output for time t. In this case we can use the input at times $t + 1$ through $t + 10$ at time t, 10 samples in the "future." Delaying the output this way is used in filters all the time.

Figure 7.1 shows a graphical way to think of convolution. Fix the time t and consider the two functions contributing to the convolution sum, x_k and h_{t-k}, as functions of k. The function h_{t-k} is a time-reversed version of the filter impulse response, with its origin at $k = t$. All we need to do to get the terms for the convolution sum is to lay this time-reversed impulse response on top of the input signal, multiply the values of x and h at each k, and add them up. The intuition that is stimulated by this picture is that the impulse response weights the past: the value h_k tells us how much weight to put on the value of the input k samples in the past.

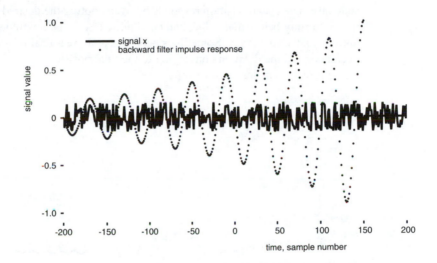

Fig. 7.1 One way to think of convolution: the time-reversed impulse response of the filter (shown dotted) weights the past of the input signal x (shown solid). The output at sample number $t = 150$ is determined by the sum of products of the overlapping signals.

8 Inverse z-transform

Let's run through the example of a reson filter being hit by a unit step digital signal. Choose the filter transfer function according to Eq. 4.2 of Chapter 5:

$$\mathcal{H}(z) = \frac{1}{1 - (2R\cos\theta)z^{-1} + R^2 z^{-2}} \tag{8.1}$$

with poles at $z = Re^{\pm j\theta}$. In fact, let's be specific and choose $R^2 = 0.97$ and $2R\cos\theta = 1.9$, so the transfer function is

$$\mathcal{H}(z) = \frac{1}{1 - 1.9z^{-1} + 0.97z^{-2}} \tag{8.2}$$

The *z*-transform of the unit step signal is just $1/(1 - z^{-1})$, so the *z*-transform of the output signal y_t is

$$\mathcal{Y}(z) = \frac{1}{(1 - 1.9z^{-1} + 0.97z^{-2})(1 - z^{-1})} \tag{8.3}$$

We can find the output signal easily enough by simply simulating it, using the update equation, Eq. 4.3 of Chapter 5,

$$y_t = x_t + 1.9y_{t-1} - 0.97y_{t-2} \tag{8.4}$$

with $x_t = 0$ for $t < 0$ and $x_t = 1$ for $t \geq 0$. The result is shown in Fig. 8.1. This is the sort of picture you're likely to find in books on control systems. It shows what happens when we try to move something with inertia from one position to another suddenly: The actual position responds by overshooting the desired final displacement, then returning below that value, and continuing to oscillate with decreasing amplitude until it finally converges. Sometimes people take a great deal of trouble to control the overshoot and the delay in converging to the final position — that's one of the jobs of the control system designer.

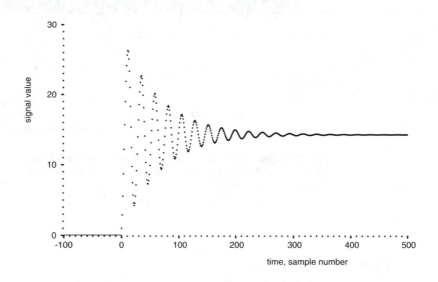

Fig. 8.1 Step response of the reson filter in the example.

Suppose we are faced with the *z*-transform of the output signal, Eq. 8.3. Is there a way we can find the corresponding output signal directly? This amounts to asking for the inverse z-transform of a given ratio of polynomials in z^{-1}. The answer is yes; there are at least three ways to do it.

Method 1: First multiply out the numerator and denominator, getting in our example

$$\mathcal{Y}(z) = \frac{1}{1 - 2.9z^{-1} + 2.87z^{-2} - 0.97z^{-3}} \tag{8.5}$$

In general, if we start with a signal that is the sum of signals like the ones shown in Table 5.1, and apply it to a feedback or feedforward filter, we will always get a ratio of polynomials in z^{-1}. Now think of the signal y_t in our example as the impulse response of a digital filter with the transfer function given by Eq. 8.5. We can implement the filter starting at $t = 0$, with input equal to the unit impulse signal, with the result:

$$
\begin{aligned}
y_0 &= 1.0 \\
y_1 &= 2.9 \\
y_2 &= 5.54 \\
y_3 &= 8.713 \quad \ldots \text{ etc.}
\end{aligned}
\tag{8.6}
$$

This checks the result plotted in Fig. 8.1, which we obtained by filtering a unit step signal.

Method 2: We can also simply divide the bottom of the ratio in Eq. 8.5 into the top, using the usual long-division algorithm. If you do this example, and think about it for a while, you'll see that this method is actually equivalent to the filtering procedure in Method 1 (see Problem 6).

Method 3: We can also proceed by guessing the general fact that the answer has one term for each pole in the z-transform. In our case, for example, there is a term corresponding to the step (a pole at $z = 1$), and terms corresponding to the poles at the complex pair of points $z = Re^{\pm j\theta}$, where R and θ are determined by $R^2 = 0.97$ and $2R\cos\theta = 1.9$. This means guessing that the z-transform for $\mathcal{Y}(z)$ can be written

$$\mathcal{Y}(z) = \frac{A}{1 - z^{-1}} + \frac{B}{1 - Re^{j\theta}z^{-1}} + \frac{C}{1 - Re^{-j\theta}z^{-1}} \tag{8.7}$$

Thus, the step response shown in Fig. 8.1 has three components — one a step function, and the other two complex exponentials — which add to an oscillatory damped waveform with the same R and θ as the original filter transfer function.

Given that the form in Eq. 8.7 is valid, we can find the inverse transform y_t by finding the inverse transform of each term; but that's easy, because the one-pole transforms are listed in Table 5.1. At this point we can simplify things by noting that the complex pole pairs must always result in real signal values. Therefore, the components of the inverse transform corresponding to the second and third terms must be complex conjugates of each other, so the imaginary parts cancel out. Thus, the coefficient C must be the complex conjugate of B. Equation 8.7 can therefore be rewritten as

$$\mathcal{Y}(z) = \frac{A}{1 - z^{-1}} + \frac{B}{1 - Re^{j\theta}z^{-1}} + \frac{B^*}{1 - Re^{-j\theta}z^{-1}} \qquad (8.8)$$

At this point it's convenient to replace $\mathcal{Y}(z)$ by its pole-zero form, and write the original and desired forms together as

$$\frac{1}{(1 - Re^{j\theta}z^{-1})(1 - Re^{-j\theta}z^{-1})(1 - z^{-1})}$$

$$= \frac{A}{1 - z^{-1}} + \frac{B}{1 - Re^{j\theta}z^{-1}} + \frac{B^*}{1 - Re^{-j\theta}z^{-1}} \qquad (8.9)$$

Note that we want this to be true for all values of z; that is, this is an *identity*. Thus, we can let z take on any value whatsoever, and this equation must still hold. We are, in fact, just about to choose some interesting values for z.

We need to find the constants A and B. This isn't hard if you do the following. Let z approach one of the poles very closely — say the pole at $z = 1$. The right-hand side of Eq. 8.9 then becomes dominated by the term corresponding to that term, and the other two terms become negligible in comparison. At the same time, the left-hand side can be thought of as two factors: $1/(1 - z^{-1})$, and the denominator of the original feedback filter, $1/(1 - (2R\cos\theta)z^{-1} + R^2z^{-2})$. As z approaches 1, nothing much happens to this latter term; it just approaches its value at the point $z = 1$, $1/(1 - 2R\cos\theta + R^2) = 1/0.07$. To summarize, the right-hand side approaches $A/(1 - z^{-1})$, and the left-hand side approaches $(1/0.07)/(1 - z^{-1})$. Therefore, $A = 1/0.07$. You can check this against the plot in Fig. 8.1. The damped exponential component corresponding to the complex pole pair almost completely dies out after a few hundred samples, and what remains is the component due to the step, with a magnitude of precisely $1/0.07 = 14.2857. \ldots$

The value of B can be found by the same procedure, and I'll leave that for Problem 7. In general, we just let z approach each of the poles in turn, and equate the dominant terms on each side of the equation. The form in Eq. 8.8, so useful for understanding the inverse z-transform, is called a *partial fraction expansion*.

I've skipped some complications that can arise in the partial fraction method. For example, we need to worry about double poles — two poles at the same point in the z-plane. In the completely general case we need to consider the case of poles repeated any finite number of times. We also need to worry about the situation when the ratio of polynomials has numerator degree equal to or greater than the denominator. (In our example the numerator degree was less than the denominator.) See Problems 11 to 13 for some discussion and work for you to do on these points.

9 Stability revisited

Let's return for a moment to the question of stability, which we first discussed in Section 2 of Chapter 5 in connection with feedback filters. The issue is whether the output of a filter grows indefinitely large with time. It's now easy to see from the partial fraction expansion (Eq. 8.7) of the z-transform that the output signal consists of a sum of terms, each corresponding to a pole in either the filter transfer function or the transform of the input signal. For the output signal to be stable, every pole in its transform must lie inside the unit circle in the z-plane. It doesn't matter whether the pole appears in the transfer function or the transform of the input signal.

In fact, we can interchange the role of input signal and filter impulse response, and get precisely the same output. Taking the example in the previous section to illustrate the point, we would get the same output if we applied the signal with z-transform $\mathcal{H}(z)$ (a damped sinusoid) to the filter with transfer function $X(z) = 1/(1 - z^{-1})$. Such a filter is a feedback filter with the defining update equation

$$y_t = x_t + y_{t-1} \tag{9.1}$$

This keeps a running total of all inputs so far, and can be called an "accumulator" or "digital integrator." Thus, the output signal plotted in Fig. 8.1 could just as well be called an "integrated damped sinusoid," instead of "the step response of a reson." This is not at all obvious intuitively, at least to me, but follows from the very simple observation that $\mathcal{Y}(z) = \mathcal{H}(z)X(z) = X(z)\mathcal{H}(z)$. Ordinary multiplication commutes; therefore, convolution commutes.

The borderline situation between stability and instability is reached when a pole occurs exactly on the unit circle. When a single pole occurs at $z = 1$, the signal stays at a constant value, neither decaying to zero nor growing indefinitely. When the pole occurs at $z = -1$ we have the situation when $R = 1$ and $\theta = \pi$ in Eq. 5.6. The signal is real, and Eq. 5.7 tells us it's just $\cos(k\pi) = (-1)^k$. That is, the signal value alternates between -1 and $+1$. You can think of this as a digital sinusoid exactly at the Nyquist frequency, the highest possible frequency. Intermediate cases occur when poles appear at points on the unit circle at angles other than 0 or π, in which cases we get sinusoids that remain at constant amplitudes forever (see Problem 5). As we see next, we can flirt with instability even more without actually going over the edge.

The observation above about interchanging input signal and filter impulse response allows us to derive an interesting z-transform. Suppose we apply a unit step signal to the digital integrator filter described by Eq. 9.1. The output signal at time t is the sum of all the input values up to and including that time, so the output is

$$y_k = \begin{cases} k + 1 & \text{if } k \geq 0 \\ 0 & \text{if } k < 0 \end{cases} \tag{9.2}$$

On the other hand, the z-transform of the output is the product of input z-transform and filter transfer function, just

$$\mathcal{Y}(z) = \frac{1}{(1 - z^{-1})^2} \tag{9.3}$$

It's a little more standard to delay this signal by one sample, multiplying the z-transform by z^{-1}, and say that the z-transform of the signal

$$y_k = \begin{cases} k & \text{if } k \geq 0 \\ 0 & \text{if } k < 0 \end{cases} \tag{9.4}$$

is

$$\mathcal{Y}(z) = \frac{z^{-1}}{(1 - z^{-1})^2} \tag{9.5}$$

Thus, a double pole on the unit circle corresponds to a signal that grows linearly with time — it's just about unstable, but not as catastrophic as the exponential growth resulting from a pole strictly outside the unit circle.

You'd be right if you guessed that the z-transform of a signal that grows as a polynomial to the nth power has a pole on the unit circle of degree $n + 1$ (see Problems 8 and 9).

To complete the picture, moving multiple poles off the unit circle in the z-plane to a radius R corresponds to multiplying the corresponding signal components by the factor R^k. We saw this in Section 5 (see Problem 10).

If you think about it, we've now covered all the cases that can arise when digital signals have z-transforms that are ratios of polynomials in z^{-1}. The signals with such transforms are always composed of sums of terms like those in Table 5.1: damped, undamped, or exploding sinusoids, possibly multiplied by polynomials in the time variable when the poles are repeated. We'll call these signals *rational-transform* signals, because a ratio of polynomials is called a *rational* function. This is precisely the class of signals that arise when impulses are applied to feedback and feedforward filters. They comprise a kind of closed universe of signals in the sense that filtering one of them results in another in the same class.

As we've seen from partial fraction expansions, the poles of a signal with a rational z-transform determine its qualitative behavior. Loosely speaking, the zeros just determine how much weight is given to each pole. Figure 9.1 shows what sort of behavior is associated with poles in different regions of the z-plane. Go over the various cases in your mind.

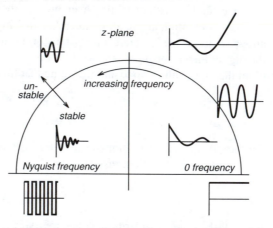

Fig. 9.1 Signal behavior associated with poles in different parts of the z-plane.

In this chapter we tied together the z used in digital filter transfer functions with the frequency content of signals. The complex symbol z^{-1} can be thought of both as a delay operator and as a frequency variable. It's now time to put our work to practical use — we're going to combine the FFT algorithm with our frequency-domain ideas to do some practical spectrum measurement. That is, we're going to see how to plot the frequency content of real signals like speech and music, This may seem relatively straightforward, but we'll need almost everything we've learned up to this point.

Notes

I avoided the nitty-gritty of partial fraction expansion, simply because we won't use it much in what follows. Similarly, I haven't developed a big table of transforms and inverse transforms. More advanced texts, like Oppenheim and Schafer's (see the Notes to Chapter 1), will fill in the details. What's important at this point is to understand the basic idea: feedback and feedforward digital filters generate the universe of rational-transform digital signals: sums of exponentially weighted sinusoids (including the special cases of polynomials and polynomial weightings). The pole locations tell the story.

In probability theory and algorithm analysis, z-transforms are called *generating functions* because the power series expansion of the z-transform "generates" the sample values. Problem 15 illustrates the application of z-transforms to a classic counting problem.

In Problem 16 I ask you to prove the very important Parseval's theorem. It shows that the z-transform preserves inner product — which underscores the fact that this frequency transform is nothing more than a change of coordinate system. As you might guess, a corresponding result holds for the DFT and other frequency transforms.

Problems

1. Why does the shape of the frequency content of the unit step signal match that of the square wave for low frequencies, but not for frequencies approaching the Nyquist?

2. Suppose we start with the digital signal f_k and define a new digital signal by interleaving zeros between its samples. That is,

$$g_k = \begin{cases} f_{k/2} & k \text{ even} \\ 0 & k \text{ odd} \end{cases}$$

Find the z-transform $G(z)$ of g_k in terms of the z-transform $\mathcal{F}(z)$ of f_k. This operation of interleaving zeros is used in oversampling, as we'll see in Chapter 14.

3. Find the z-transform of the signal

$$f_k = \begin{cases} 1/k! & \text{if } k \geq 0 \\ 0 & \text{if } k < 0 \end{cases}$$

Recall that $0! = 1$.

4. Equation 6.6 in Chapter 5 gave the following as the impulse response of a feedback digital filter:

$$y_t = R^t \left[\frac{\sin(\theta(t+1))}{\sin\theta} \right]$$

What is the z-transform of y_t?

5. When $R = 1$, the impulse response of the reson given above becomes a pure sine wave, except for a constant factor. Thus, we can generate a sine wave by implementing the reson feedback filter with a unit impulse input. Is this a practical way to generate sine waves? What might go wrong? Try it.

6. Prove that the long-division method of inverting a z-transform, Method 2 in Section 8, is equivalent to the impulse-response method, Method 1.

7. Find the value of B in the partial fraction expansion of Eq. 8.8. Then find the output signal y_t in the form of a step function plus a damped sinusoid.

8. Derive the z-transform of the linearly increasing signal in Eq. 9.4 by differentiating the z-transform of a unit step signal term by term. What results can you get by repeating this procedure?

9. Use what you learned from Problem 8 to find the z-transform of the signal $y_k = k^2$, $k \geq 0$.

10. In Section 5, we derived new z-transforms by weighting the signal at time t by the factor R^t. Apply this idea to find the z-transform of the signal $y_k = k R^k \sin k\theta$, $k \geq 0$.

11. When a ratio of polynomials in z^{-1} has a double pole at $z = p$, the partial fraction expansion must include the terms

$$\frac{A}{1 - pz^{-1}} + \frac{B}{(1 - pz^{-1})^2}$$

Devise a procedure for finding the coefficients A and B.

12. Suppose we want to expand a ratio of polynomials in z^{-1} in which the degree of the numerator is not smaller than that of the denominator. For example,

$$\frac{a_0 + a_1 z^{-1} + a_2 z^{-2} + a_3 z^{-3} + a_4 z^{-4}}{(1 - p_1 z^{-1})(1 - p_2 z^{-1})}$$

Explain how we can get a partial fraction expansion with the usual terms corresponding to the poles, namely

$$\frac{A}{1 - p_1 z^{-1}} + \frac{B}{1 - p_2 z^{-1}}$$

13. You might think of the following mathematical trick to handle cases like a repeated pole. Start with two distinct poles, say p_1 and p_2, use the simple partial fraction expansion for distinct poles, and then take the limit as $p_1 \to p_2$. Can you derive the result of Problem 11 this way?

14. The digital signal in Problem 3 is not a rational-transform signal. Find other examples of signals with non-rational transforms..

15. Here's a classic mathematical problem that goes back to Leonardo Fibonacci (? – *ca* 1250). To get a neat formulation we're going to make the extreme assumptions that every pair of rabbits matures in one month, and produces a pair of baby rabbits the month after reaching maturity and every month thereafter. Start with one pair of baby rabbits at the beginning of Month 0. At the beginning of Month 1 this pair matures, but there will still be only one pair of rabbits. By the beginning of Month 2, however, there will be two pairs: the original pair, plus one new baby pair born to that original pair. By the beginning of Month 3, there will be only one more pair, for a total of three pairs, because the baby pair is not yet able to reproduce. By the beginning of Month 4, however, there will be a total of five pairs, three from the preceding month, plus two more born to the pairs that were mature that preceding month.

Denote the number of pairs of rabbits at the beginning of Month t by r_t.

(a) Derive an expression for r_t in terms of r_{t-1} and r_{t-2}. (We already know that $r_0 = 1, r_1 = 1, r_2 = 2, r_3 = 3$, and $r_4 = 5$.)

(b) Interpret r_t as the output signal of an appropriate digital filter with appropriate initial conditions, the values of the input and output signals at the beginning of its operation. Is the filter feedback or feedforward? Is it stable?

(c) Find $\mathcal{R}(z)$, the z-transform of r_t.

(d) Find the poles and the corresponding partial fraction expansion of $\mathcal{R}(z)$.

(e) Find an explicit expression for r_t by taking the inverse z-transform of the partial fraction expansion.

(f) How many pairs of rabbits will there be after one year?

16. Let $X(\omega)$ and $Y(\omega)$ be the z-transforms evaluated on the unit circle of digital signals x and y, respectively. Define the inner product of X and Y by

$$\langle X, Y \rangle = \frac{1}{2\pi} \int_{-\pi}^{\pi} X(\omega) Y^*(\omega) \, d\omega$$

Prove *Parseval's theorem*:

$$\langle X, Y \rangle = \langle x, y \rangle$$

where the inner product between signals x and y is defined as in Eq. 2.3. Choose some particular signals x and y for which this theorem is easy to verify, and then verify it. What does the result mean when $x = y$?

CHAPTER 10

Using the FFT

1 Switching signals on

The wonderful FFT algorithm for computing the Discrete Fourier Transform (DFT) revolutionized signal processing. It is a very efficient way to get information about the frequency content of signals — both real-world and artificially generated. If it's used with proper understanding it can be a friendly, helpful companion. But it's important to be aware of its limitations. There are snares for the unwary.

We saw an example of the kind of problem to expect at the end of Chapter 8, where we computed the DFT of a single phasor. The DFT worked fine when the frequency of the phasor coincided with one of the sample points on the unit circle, say point 133 out of 1024 points. But when we looked at the DFT of a phasor with frequency corresponding to 133.5/1024 times the sampling frequency, a frequency in the "crack" between the DFT points, we got rather disappointing results. The computed spectrum couldn't tell us there was precisely one frequency component present; instead, it showed a wide distribution of many DFT frequencies near points 133 and 134. (See Figs. 9.1 and 9.2 in Chapter 8 again.)

Before we look more closely at the FFT, I want to clear up a possible source of confusion. We often use complex phasors instead of sines and cosines because the algebra is simpler. In practice, though, we usually use the FFT on real-valued signals. As pointed out in Chapter 8, the frequency content of real-valued signals is an even function of frequency. In the case of real-valued signals the frequency points above the Nyquist frequency are redundant, and there's no reason to plot them. However, I'll continue to use examples with complex phasors for algebraic simplicity, plotting the entire range of frequencies from zero to the sampling frequency, or sometimes, when it's more convenient, from minus the Nyquist frequency to plus the Nyquist.

Let's reexamine the example at the end of Chapter 8 in the light of what we've learned about the z-transform. In particular, let's take another look at the z-transform of a phasor. In the previous chapter, we always considered signals that start at $t = 0$. But what happens when a signal doesn't start at a particular time, but has been present forever (is *two-sided*)? The z-transform of the two-sided phasor with frequency θ radians per sample,

$$x_t = e^{j\theta t}, \qquad -\infty \leq t \leq \infty \tag{1.1}$$

is

$$X(z) = \sum_{t=-\infty}^{\infty} e^{j\theta t} z^{-t} \tag{1.2}$$

This is a slightly disguised form of a sum we've seen before. To put it in a more familiar form, evaluate it on the unit circle, yielding the frequency content of the two-sided phasor x_t. Setting $z = e^{j\omega}$, Eq. 1.2 becomes

$$X(\omega) = \sum_{t=-\infty}^{\infty} e^{j(\theta-\omega)t} \tag{1.3}$$

This is exactly the same as the left-hand side of Eq. 3.3 in Chapter 9, which we used to establish the orthogonality of the basis for z-transforms. It's just a sequence of δ functions, spaced at intervals of 2π, with the independent variable $\theta - \omega$. The periodicity of 2π is irrelevant here, since the function is defined on the circle in the z-plane. The frequency content therefore has a single δ function on the unit circle at $\omega = \theta$, as shown on the left in Fig. 1.1. This makes perfect sense — there is only one frequency present, θ, and the frequency content must be zero everywhere else. Put another way, the complex phasor is an element of the basis used to develop the z-transform in Section 2 of Chapter 9.

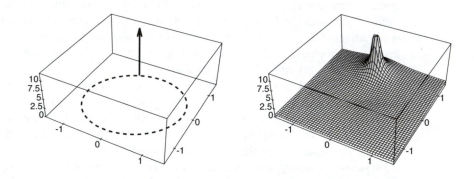

Fig. 1.1 On the left, the frequency content of a complex phasor at a single frequency, a δ function on the unit circle; on the right, the frequency content when the phasor is turned on abruptly at time 0.

Now return to the one-sided case. We've already derived the z-transform of the one-sided phasor $e^{j\theta t}u_t$. Just set $R = 1$ in Eq. 5.6 of Chapter 9:

$$\frac{1}{1 - e^{j\theta}z^{-1}} \tag{1.4}$$

As usual, we evaluate the magnitude of this on the unit circle in the z-plane to get the frequency content of the one-sided phasor, plotted in Fig. 1.2. A contour plot above the z-plane is also shown on the right in Fig. 1.1, to contrast with the δ function when the phasor is present for all time. The abrupt switch of the signal from perfectly zero to the sinusoid has introduced some contributions from all frequencies. The frequency content still has an infinite value precisely at the frequency θ, but it also rises up to infinity in the neighborhood of θ. Suddenly turning on the signal makes the spectrum less definitely localized at one particular frequency.

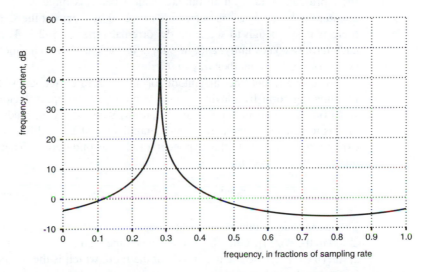

Fig. 1.2 Frequency content of a one-sided digital phasor with frequency $\theta = 9/32 = 0.28125$ times the sampling rate. The peak at this point theoretically goes to infinity.

The frequency spreading of the chopped phasor is consistent with what we've learned already about spectra in general. Any kind of abrupt jump in a signal generates a broad range of frequencies. This accounts for the clicks and pops in carelessly edited digital audio — recall the slow decay of the spectrum of a square wave, Fig. 3.4 in Chapter 7; and the broad frequency content of the unit step signal, Fig. 4.3 in Chapter 9.

2 Switching signals on and off

What happens if we now switch the phasor back off after a given number of samples, say n? The z-transform of the resulting finite stretch of a phasor is

$$1 + e^{j\theta}z^{-1} + e^{j2\theta}z^{-2} + \cdots + e^{j(n-1)\theta}z^{-(n-1)} \tag{2.1}$$

This is a finite geometric series, each term being the previous multiplied by $e^{j\theta}z^{-1}$. It isn't hard to show that the closed form is:

$$\frac{1 - e^{jn\theta}z^{-n}}{1 - e^{j\theta}z^{-1}} \tag{2.2}$$

(see Problem 1). The numerator is just one minus the term that would be next. It's worth looking at this result in some detail.

Notice first that this z-transform is, after all, the sum of a finite number of terms. It therefore cannot take on an infinite value when z is finite and θ is a real angle. The apparent pole at $z = e^{j\theta}$ is bogus; it's canceled by a zero at the same point. The value there is easy enough to see from the original series, Eq. 2.1. The factor multiplying each term becomes unity, and the sum is just one added up n times, and thus equals n (see Problem 2 for another way to see this).

Next, we'll put the magnitude of Eq. 2.2, the frequency content, in a very illuminating form by playing around a little with complex factors. The trick is the same one we used in Eq. 7.6 of Chapter 4, when studying feedforward filters. The idea is to rewrite the numerator and denominator of Eq. 2.2 by factoring out complex exponentials with half the exponent of the second terms. The numerator, with z replaced by $e^{j\omega}$, becomes

$$1 - e^{jn(\theta-\omega)} = e^{jn(\theta-\omega)/2}\left[e^{-jn(\theta-\omega)/2} - e^{jn(\theta-\omega)/2}\right] \tag{2.3}$$

The magnitude of this becomes the magnitude of the factor in brackets, which is just $2|\sin(n(\theta-\omega)/2)|$. In the same way, the magnitude of the denominator is $2|\sin((\theta-\omega)/2)|$, so the magnitude of the ratio, which is the frequency content of the finite stretch of phasor, is

$$\left| \frac{\sin(n(\theta-\omega)/2)}{\sin((\theta-\omega)/2)} \right| \tag{2.4}$$

Figure 2.1 shows a plot of this for $n = 32$ samples of a phasor of frequency $\theta = 9/32$ times the sampling rate, together with the frequency content of the one-sided phasor at the same frequency. This looks like a mess, but it's not just any mess. It tells us, after all, exactly how much of each frequency we need to get a sinusoid that starts abruptly at $t = 0$ and ends abruptly at $t = n-1$. If you look at it that way, it's a miracle we can figure this out at all. The peak does occur at the expected frequency $\omega = \theta$ (see Problem 3), and the general shape does follow the frequency content for the one-sided case. We've gone from a single, ideal δ function for a two-sided phasor, to a smooth curve with an infinite peak for a one-sided phasor, to this oscillatory, finite-valued curve for the n-sample phasor.

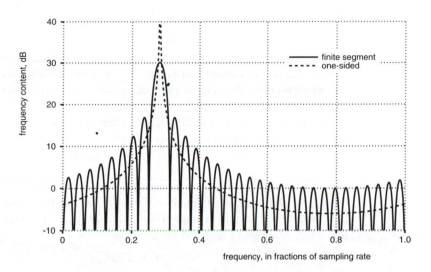

Fig. 2.1 The frequency content of a finite segment of a digital phasor at frequency θ = 9/32 = 0.28125 times the sampling rate. The segment is 32 samples long. For comparison, the dashed line shows the frequency content of the one-sided phasor at the same frequency, from Fig. 1.2.

The oscillations in frequency content are actually easy to predict from the z-transform, Eq. 2.2. The numerator has *n* zeros equally spaced on the frequency axis, and all but the one canceled at the peak frequency contribute nulls to this curve.

Now that we are turning the signal off as well as on, we get even more spreading of the spectrum.

3 Resolution

The phenomenon we've just seen comes up in many fields of science. In astronomy, the issue is usually couched in terms of the resolving power of a telescope, its ability to separate the image of two stars close together. Everyone who uses telescopes knows that as the aperture of a telescope widens, its resolving power increases. Exactly the same principle applies to measuring the spectrum of signals with the DFT. The ability to distinguish between two audio tones that are close in frequency improves as the record length increases.

The astronomical and the audio examples are closer than you might think. Mathematically they are identical except that the optical case is continuous and two-dimensional, while the audio case is discrete and one-dimensional. The DFT terminology reflects the analogy. We say that we are looking at the phasor through a *window* that is *n* samples wide. In this section I want to demonstrate directly that wider windows mean finer frequency resolution.

Selecting n consecutive samples of a signal amounts to using a *rectangular* window w_t. That is, if we start with the infinite-extent signal x_t, the windowed version of x_t is

$$y_t = w_t x_t \tag{3.1}$$

for all t, where the window function w_t is a constant inside some finite range of time values, and is zero outside that range. As we'll see shortly, there are good reasons to use windows other than the simple rectangular one, so let's think of the window function as having some general shape given by w_t.

Return to the example of the spectrum measurement of a phasor and substitute the phasor at frequency θ for x_t in Eq. 3.1 to get

$$y_t = w_t e^{j\theta t} \tag{3.2}$$

We encountered this relationship between two signals before, in Chapter 9. We saw there from the defining summation of the z-transform that these signals' z-transforms are related by a simple change of variable. That is, the z-transform of y_t is the z-transform of w_t with z replaced by $ze^{-j\theta}$:

$$\mathcal{Y}(z) = \mathcal{W}(ze^{-j\theta}) \tag{3.3}$$

In terms of the frequency variable ω, z is $e^{j\omega}$, and therefore this tells us that the frequency content of y_t is just the frequency content of w_t shifted by θ :

$$Y(\omega) = W(\omega - \theta) \tag{3.4}$$

To study the effect of a window on a phasor, then, we might as well take $\theta = 0$. The spectrum shaping caused by windowing a phasor of any other frequency will be the same, but shifted by θ.

One final point before we look at the effect of window length on the frequency content of a windowed phasor. The value $W(0)$ is the measured value of frequency content precisely at $\omega = 0$. In terms of a telescope, this is the brightness of the image at the true star position. If we want to compare two windows, it's reasonable to adjust the multiplicative scale so that the values of $W(0)$ for the two windows are equal. This is especially easy to arrange because $W(0)$ is simply

$$W(0) = \mathcal{W}(1) = \sum_{t=0}^{n-1} w_t \tag{3.5}$$

using the defining z-transform summation. Thus, we'll normalize windows by choosing

$$\sum_{t=0}^{n-1} w_t = 1 \tag{3.6}$$

The n-point rectangular window normalized to make $W(0) = 1$ is then given by

$$w_t = \begin{cases} 1/n & \text{if } 0 \le t < n \\ 0 & \text{else} \end{cases} \tag{3.7}$$

We'll bother to normalize windows this way only when we're comparing them. When we actually use them, it's usually simpler not to.

Fig. 3.1 shows a comparison of the frequency content of two rectangular windows, for lengths $n = 8$ and 64 samples. Bear in mind that this shows the spreading of a single-frequency signal caused by looking at only a finite stretch of it. Given the simplicity of this operation, the result is a rather spectacular splatter. The improved resolution of the 64-point window is quite clear. Not only does its frequency content have a narrower peak than the 8-point window at $\omega = 0$, but its values at other frequencies, called its *side lobes*, fall off faster. These side lobes play a critical role in determining how good a window is, because they show the extent to which the observed central frequency "leaks" to neighboring frequencies. Reciprocally, they show the extent to which the components at neighboring frequencies leak to the region near the central frequency.

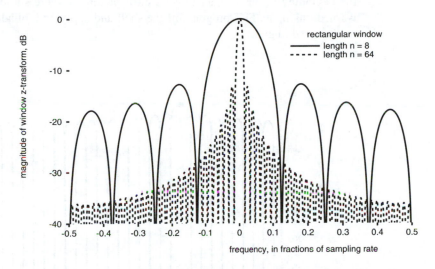

Fig. 3.1 Comparison of the frequency content of two rectangular windows, with lengths $n = 8$ and 64 samples.

One measure of resolution quality is the width of the central lobe. In practical optical situations, for example, two stars close together can be distinguished only if the central lobes of their images do not overlap much. The width of the central lobe, zero-crossing to zero-crossing, is determined by the fact that the central peak is straddled by two zeros in the numerator of the window z-transform, Eq. 2.2. The zeros of an n-point rectangular window are equally spaced around the circle, so they're separated by $2\pi/n$ radians, or, in terms of the sampling rate, f_s/n Hz. The width of the central lobe is therefore $2f_s/n$ Hz. Notice the important fact that as n increases, the spacing between two resolvable frequencies decreases as $1/n$.

In summary, two features of a window's frequency content affect the spreading of the energy of the original signal: the width of the central lobe, and the height of the side lobes.

Before we go ahead to the DFT, I'd like to go back and savor Eq. 3.4. It applies generally and says that multiplying a signal by a phasor of frequency θ shifts the signal's frequency content by θ. The principle is used constantly in radio and television receivers, where it's called *heterodyning*. For example, most household AM radio receivers heterodyne all incoming-station center frequencies to a standard 455 kHz, called the *intermediate frequency*, or IF.

4 The DFT of a finite stretch of phasor

We're finally in a position to understand what happens when we take the DFT of a stretch of phasor at a frequency that lies in the crack between the DFT points. This is where we left off at the end of Chapter 8. We needed all the work in Chapter 9 and at the beginning of this chapter to get a clear picture — there's a lot going on. As I've warned, taking an FFT program off the shelf and applying it blindly to a piece of signal can lead to grief.

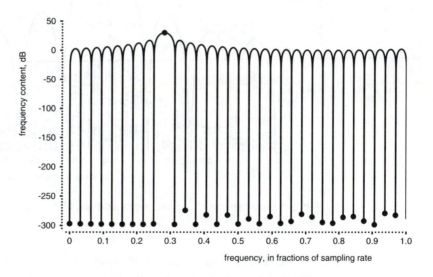

Fig. 4.1 The 32-point FFT of a finite stretch of phasor at frequency 9/32 times the sampling frequency, superimposed on the frequency content before sampling in the frequency domain. The points down around −300 dB are not zero because of roundoff error in the FFT computation.

In fact, the DFT is nothing more than the *z*-transform of *n* consecutive samples of a signal, evaluated at *n* equally spaced frequency points. Thus, the DFT of the *n* samples of the phasor we've been considering is obtained by evaluating the expression in Eq. 2.2 at the DFT points; the resulting magnitude values are shown by the exaggeratedly large dots in Fig. 4.1. The value of the DFT at the point corresponding to the frequency of the phasor, $\theta = 9/32$ times the sampling rate, is $n = 32$, or about 31 dB. The vertical scale in that graph has been expanded down to -300 dB so we can see the $n - 1$ points that are not at the frequency θ. These values are theoretically zero, but because of the numerical roundoff noise in the FFT computation, they turn out to be zero to within about 15 decimal places. (I used double-precision floating-point arithmetic, 64 bits.) This situation, where the frequency of the phasor is precisely equal to a DFT point, is analogous to the example shown in Fig. 9.1 of Chapter 8 for 1024 points.

The situation shown in Fig. 4.1 is a fluke; it's extremely unlikely that a measured signal will have a frequency that is so nicely related to the sampling frequency and the size of the DFT. Besides, signals usually are not single phasors at all, but some conglomeration of many frequencies, often moving around. Figure 4.2 shows an example of a more common situation; in this case the phasor frequency is $\theta = 9.333/32$ times the sampling rate. We saw this kind of phenomenon back in Fig. 9.2 of Chapter 8. The DFT points are not lined up with the nulls in the spectrum, and we get a much more accurate picture of just how dispersed the frequency content is around the true frequency of the phasor. This example also reminds us that a true signal frequency will almost always fall in the cracks between the DFT points, and will almost never be precisely equal to the DFT point with the highest measured value.

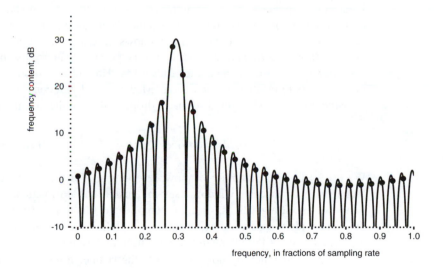

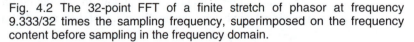

Fig. 4.2 The 32-point FFT of a finite stretch of phasor at frequency 9.333/32 times the sampling frequency, superimposed on the frequency content before sampling in the frequency domain.

To summarize, taking the DFT of a finite stretch of a digital phasor introduces uncertainty about its true frequency θ in the following three ways, the first two resulting from the windowing process, and the third resulting from the process of sampling in the frequency domain:

> (a) energy is spread around frequencies close to θ within the main lobe of the window's transform;
> (b) energy is spread to frequencies far from θ because of the side lobes of the window's transform;
> (c) uncertainty is introduced in actual frequency location because the frequency content of the windowed signal is computed only at the n DFT points.

If we go back to some original analog signal that has been sampled, we should add another source of problems:

> (d) false frequencies are introduced by aliasing in the original sampling process.

Given all these difficulties, you can understand my earlier expression of caution.

On the positive side, it turns out that we can lower the side lobes significantly by using nonrectangular windows — but as we'll see next, not without a certain price.

5 The Hamming window

If turning a signal on and off abruptly causes problems, then it should improve things to use more gradual transitions. This simple observation is actually a great idea, and a large repertoire of elegant and useful windows have been invented over the years. Figure 5.1 shows the *Hamming* window, a very popular compromise between simplicity and effectiveness. It's named after Richard W. Hamming, a pioneer in the application of computers to practical computation. Mathematically, it consists of a single cycle of a cosine, raised and weighted so that it drops to 0.08 at the end-points and has a peak value of one:

$$h_t = 0.54 - 0.46\cos(2\pi t/(n-1)), \qquad 0 \le t < n \qquad (5.1)$$

We use h_t for the Hamming window, reserving w_t for the rectangular window. The fact that this particular window uses the cosine has no magical significance. Other windows use straight lines, or more complicated functions. But its presence here, in this very widely used window, is particularly felicitous, because it makes the analysis especially easy. After all, sinusoids have been our friends since Page 1.

I hope your reflex by now is to write the cosine in terms of phasors. Remember that there is an implicit rectangular window in the definition of Eq. 5.1, because $h_t = 0$ outside the indicated range. We can therefore rewrite Eq. 5.1 as

$$h_t = \left[0.54 - 0.23e^{j2\pi t/(n-1)} - 0.23e^{-j2\pi t/(n-1)} \right] w_t \qquad (5.2)$$

where w_t is the rectangular window. The first term is just a copy of the rectangular window, and the last two are just heterodyned versions. (Remember Eqs. 3.2 and 3.4:

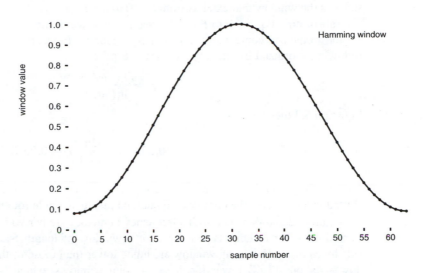

Fig. 5.1 The 64-point Hamming window.

multiplying a time function by a phasor shifts its frequency content.) The frequency content of the Hamming window, which we'll denote by $H(\omega)$, is therefore

$$H(\omega) = 0.54 W(\omega) - 0.23 W(\omega - \frac{2\pi}{n-1}) - 0.23 W(\omega + \frac{2\pi}{n-1}) \qquad (5.3)$$

So the transform of the Hamming window is a sum of shifted and weighted versions of the transform of the rectangular window. Intuitively, the idea is to arrange the shifts and weights to cancel adjacent side lobes.

Now let's get to work on the transform of the rectangular window, $W(\omega)$. We found its magnitude in Eq. 2.4, but now it's important to preserve the phase information, precisely because we're counting on canceling pieces of $W(\omega)$ when we shift and add. Go back to Eq. 2.2, set $\theta = 0$ (since it doesn't matter what phasor we're analyzing), and set $z = e^{j\omega}$ (as usual), to get

$$W(\omega) = \frac{1 - e^{jn\omega}}{1 - e^{j\omega}} \qquad (5.4)$$

The next operation should also be familiar by now. As shown in Eq. 2.3, factor out a complex exponential in both the numerator and denominator with half the angle of the complex exponential already there, yielding

$$W(\omega) = e^{-j(n-1)\omega/2} \frac{\sin(n\omega/2)}{\sin(\omega/2)} \qquad (5.5)$$

This form has a simple interpretation. The complex exponential in front represents a shift of the window $(n-1)/2$ samples to the right, which moves it so it extends from 0 to $n-1$, instead of being centered at 0. The remaining factor is the frequency content of the centered window (the ''zero-phase'' version). Of course its magnitude is the same as the window that extends from $t = 0$ to $n-1$.

Substituting Eq. 5.5 into Eq. 5.3 gives us what we're after, an explicit expression for the frequency content of the Hamming window. First, it's convenient to use the following shorthand for the ratio of sines in Eq. 5.5:

$$S(\omega) = \frac{\sin(n\omega/2)}{\sin(\omega/2)} \tag{5.6}$$

Equation 5.3 then becomes

$$H(\omega) = e^{-j(n-1)\omega/2}\left[0.54S(\omega) + 0.23S(\omega - \frac{2\pi}{n-1}) + 0.23S(\omega + \frac{2\pi}{n-1}) \right] \tag{5.7}$$

The plus signs in front of the second two terms are not misprints; substituting the shifted values of ω in the complex exponential results in angle rotations by π.

Figure 5.2 shows a plot of this frequency content, together with the corresponding transform of the rectangular window, for a window of length 64. As promised, the side lobes of the Hamming window are much lower than those of the rectangular window — about 17 dB lower for these 64-point windows, which is a factor of about seven.

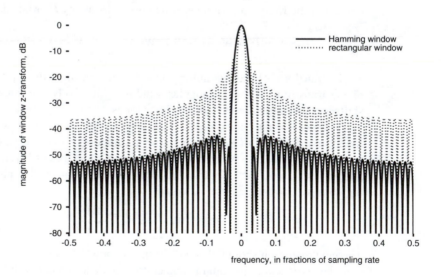

Fig. 5.2 Comparison of frequency content of 64-point Hamming and rectangular windows.

The price for lower side lobes is a broadened central lobe. This seems unavoidable if we tamper with the shape of the transform by shifting and adding. It seems inevitable that trying to cancel adjacent side lobes causes reenforcement at the central lobe. Figure 5.3 confirms this, showing a close-up of the frequency content of the Hamming and rectangular windows at frequencies near zero. The rectangular window has its first null, caused by a zero directly on the unit circle, at $\omega = \pi/64 = 0.015625\pi$, while the Hamming window has its first, and rather more shallow, null at $\omega = 0.0322\pi$. The central lobe is thus about twice as wide. (The effect is not nearly so bad for the Hamming window as it might seem at first, because the second lobe of the rectangular window comes back up, to almost -13 dB, very quickly.) This trade-off between resolution at the central frequency and leakage from components at neighboring frequencies is an unavoidable law of nature. The art of window design is to get the lowest side lobes for a given resolution, or vice versa.

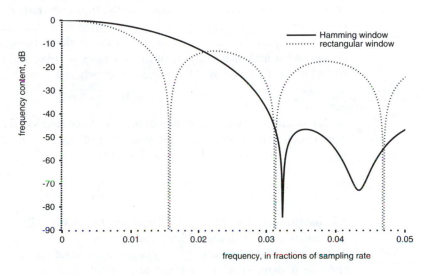

Fig. 5.3 Close-up of the previous figure, comparing the resolution of the 64-point Hamming and rectangular windows.

6 Windowing in general

Up to now, we've considered the effect of windowing only on a single phasor. But we know that any signal can be broken down into phasors, and that means we can find the effect of a window on completely arbitrary signals with very little additional work.

Suppose then that we start with an arbitrary digital signal x_t, which is written in terms of its frequency content $X(\theta)$ as

$$x_t = \frac{1}{2\pi} \int_{-\pi}^{\pi} X(\theta) e^{j\theta t} \, d\theta \tag{6.1}$$

This is the inverse z-transform, Eq. 2.2 in Chapter 9. We know that windowing each phasor component $e^{j\theta t}$ converts the phasor to a signal with frequency content $W(\omega - \theta)$, a shifted version of the window transform (recall Eq. 3.4). Therefore, replacing the phasor in the integrand by $W(\omega-\theta)$ yields the transform $Y(\omega)$ of the windowed version of the original signal x_t:

$$Y(\omega) \;=\; \frac{1}{2\pi} \int_{-\pi}^{\pi} X(\theta)\,W(\omega-\theta)\;d\theta \tag{6.2}$$

Equation 6.2 is a remarkable result in itself and goes far beyond the context of windowing for spectrum measurement. Think about its significance. The integral combines $X(\omega)$ and $W(\omega)$ to obtain $Y(\omega)$. Where have you seen an operation like this before?

The integral in Eq. 6.2 is the *convolution* of the two frequency functions $X(\omega)$ and $W(\omega)$. The convolution is with respect to the frequency rather than the time variable, the summation in the case of filtering in the time domain is replaced by integration, and the independent variable θ ranges over the circle — but those differences are all details. What matters is that one of the functions is being slid past the other and the result tallied as a function of the displacement between them. We can write Eq. 6.2 as

$$Y(\omega) \;=\; W(\omega) * X(\omega) \tag{6.3}$$

It's worth comparing these two examples of convolution side-by-side. Here's the convolution that embodies the digital filtering of signal u_t by the filter with impulse response h_t to produce v_t, Eq. 7.3 in Chapter 9:

$$v_t \;=\; u_t * h_t \;=\; \sum_{k=-\infty}^{\infty} u_k h_{t-k} \tag{6.4}$$

The independent variable can be frequency (as in Eq. 6.2) or time (as in Eq. 6.4), continuous (6.2) or discrete (6.4), and its range can be finite (6.2) or infinite (6.4). The result will always have the meaning of a convolution, and the counterpart in the opposite domain (time for a frequency variable and vice versa) will be point-by-point multiplication. Thus, Eq. 6.4 corresponds to the frequency-domain relation $V(\omega) = H(\omega)\,U(\omega)$, point-by-point multiplication in the opposite domain.

The interpretation of windowing in the time domain as filtering in the frequency domain is a good way to see what the DFT does when applied to a windowed segment of a signal. The frequency content of the signal (not the signal itself) is convolved with the frequency transform of the window shape. The closer the window transform is to an ideal pulse — the narrower the first lobe and the lower the side lobes — the closer the result will be to the original signal spectrum.

That point-by-point multiplication in one domain corresponds to convolution in the other is a striking example of symmetry between the time and frequency domains. In a way it cuts in half the things you need to remember, and at least doubles your intuition.

7 Spectrograms

As we've seen, the more points in a segment of a sinusoid, the better the frequency resolution. You might think that in general, the longer the segment used for the DFT of a signal, the better off we are in all respects. Life is not so simple, however, and there is a price to pay in using very long windows. The problem has to do with resolution in the *time* domain. Interesting signals are never sums of sinusoids that continue for long periods of time. They change, often rapidly and in complicated ways, and that's what makes them interesting. Trying to break such signals down into periodic basis elements with the DFT can result in gross misrepresentations of what's actually happening. The DFT and its fast FFT algorithm are handy tools, but they color the way we see signals and are easy to misuse.

To illustrate the trade-off between time and frequency resolution, we'll look at the analysis of a real signal, a typical phrase from the call of the northern cardinal. Sounds we're used to hearing, like bird calls, music, or speech, often have reasonably stable frequency content over time intervals of the order of 1/120 sec. That corresponds to about 184 samples at a sampling rate of 22.05 kHz, or 368 samples at 44.1 kHz, two standard sampling rates used for digital sound. The FFTs usually used for sound are therefore powers of two in this ballpark, usually not shorter than 128 samples or longer than 1024 samples. What we ordinarily do, then, is compute a sequence of n-point FFTs, starting with a window at the first n samples of the signal, and then slide over some amount for each successive FFT. Figure 7.1 shows the scheme with a window length of 1024 samples and a sliding increment of 200 samples. With these numbers, each second of sound at 22.05 kHz samples per sec yields 111 FFTs, the first using points 0 through 1023 and the last using points 22,000 through 23,023 (and thus extending past the one-second mark).

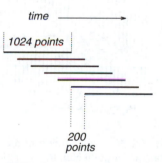

Fig. 7.1 Computing the FFT of a signal at sliding windows. Each window in this example has length 1024 samples, and the slide increment is 200 samples.

We can visualize all the data we get from the sequence of sliding FFTs in the *spectrogram*, which we've shown before without much of an explanation. The frequency content is now actually a function of two variables: frequency, which we use as the ordinate, and time, which we use as the abscissa. The magnitude of the frequency content is then indicated by gray level — the greater the magnitude, the darker the ink.

Figure 7.2 shows the cardinal call displayed this way, with an FFT window size of 1024 samples, and a time increment from each window to the next of 33 samples. This spectrogram suggests a qualitative description of the sound that jibes well with the actual perception, a slurred whistle with a break of some sort in the middle:

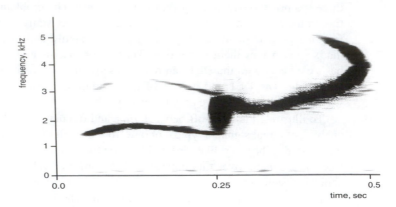

Fig. 7.2 Spectrogram of a call of a male northern cardinal. The abscissa is time in sec, and the ordinate is frequency in kHz. The FFT used a Hamming window with length 1024 samples, the time increment from slice to slice was 33 samples, and the sampling rate was 22.05 kHz.

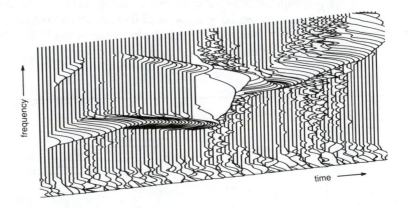

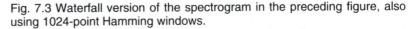

Fig. 7.3 Waterfall version of the spectrogram in the preceding figure, also using 1024-point Hamming windows.

(a) A note with well-defined pitch bends up and then down. The note has a clear second harmonic.

(b) At about 0.2 sec, virtually all the energy is transferred to the second harmonic, and the pitch trajectory continues down and then up.

(c) A component at a higher pitch begins at about 0.36 sec and descends to meet the note at the end, which occurs at about 0.46 sec.

The jump at about 0.2 sec is a phenomenon we might want to study in detail, but the event appears to be quite blurry. There's an interval from about 0.19 to 0.24 sec where a broad band of energy extends between the first and second harmonics, and then only the second harmonic emerges.

Figure 7.3 shows an alternative visualization of the same data, called a *waterfall* plot. The attempt here is to give the illusion of a three-dimensional plot, with distance from the time-frequency plane representing frequency content. If you have trouble perceiving it that way, rotate the picture $+90°$; for some reason the human eye likes to see mountains.

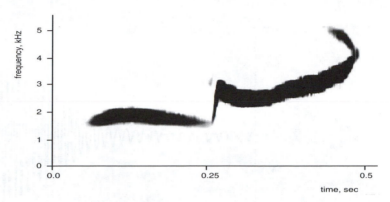

Fig. 7.4 Same as Fig. 7.2, except the window length was 128 samples.

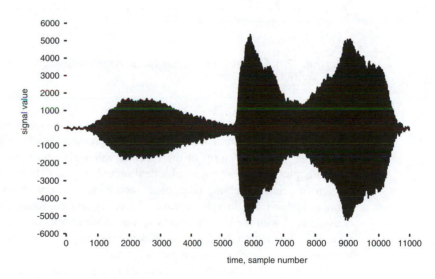

Fig. 7.5 Time waveform of the same cardinal call.

Figure 7.4 shows the spectrogram corresponding to the same sample of sound, also with a Hamming window, but with a window length of only 128 samples. The transition is now revealed as a definite jump, with much clearer starting and ending points. There's even a trace of the jump in the second harmonic. It's fair to say that the time resolution is improved over the 1024-point analysis. That's easily understandable, because each FFT now reflects the events in the span of only 128 points, so there's less averaging of the signal characteristics at each abscissa. On the other hand, the bands indicating the pitch are broader, which means we have a less accurate measurement of the pitch at any given time. We already expect this from our discussion of resolution in Section 3.

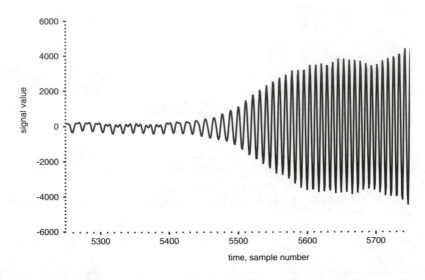

Fig. 7.6 Close-up of the previous waveform, showing the time region where the pitch jumps.

The frequency content displayed in a spectrogram is usually normalized so that the effects of signal amplitude fluctuations are suppressed. To illustrate this point, Fig. 7.5 shows the actual waveform of the cardinal call we've been analyzing. You can see from this plot that the signal amplitude diminishes gradually up to the point where the sudden frequency-doubling takes place, and then increases sharply at that point. This is not readable from the spectrograms. To verify that the sudden increase in amplitude is associated with the shift in energy to the second harmonic, Fig. 7.6 shows a close-up of this time region. Notice how suddenly the shape of the waveform changes — the transition occurs in about 40 samples, or 2 msec.

Finally, Fig. 7.7 shows the effect of using a 128-point rectangular window, instead of the 128-point Hamming window, in the spectrogram of Fig. 7.4. The leakage of energy outside the central lobe smears the spectral information appallingly. Our work on windows was well worth the effort. (See Problem 10 for a question about this picture.)

In this chapter we've looked at some practical problems in using the FFT, one of the most commonly used signal-processing algorithms. If you consider everything that happens when you just take a piece of a sine wave and plug it into an FFT program, I think you'll agree that the subject has its subtle side. To understand exactly what comes out of the FFT, we used two frequency transforms — the z-transform and DFT — and the idea of windowing.

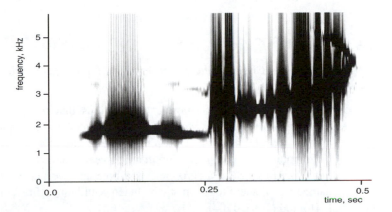

Fig. 7.7 Same as Fig. 7.4 (128-point Hamming windows), except a rectangular window was used.

Notes

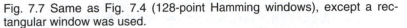

The analogies between Fourier analysis and imaging in telescopes are so accurate and striking that I couldn't resist mentioning some of them in Section 3. Windowing in optical systems is called *apodization*, from the Greek meaning "without feet" — because the side lobes of a star image look like little feet. Apodization is widely used in optics for exactly the same reasons we've found it useful here. For more, see the following review article (note the British spelling):

> P. Jacquinot and B. Roizen-Dorrier, "Apodisation," in *Progress in Optics*, vol. III, E. Wolf (ed.), North Holland Publishing, Amsterdam, 1964.

I've also seen the term *aperture shading*, which is nicely descriptive.

Some common alternatives to the Hamming window are defined in Problems 6 to 8. There are also fancier windows, such as the Kaiser and Dolph-Chebyshev windows, which allow the user to trade off the width of the central lobe with the height of the side lobes. For example, see

> T. Saramäki, "Finite Impulse Response Filter Design," in *Handbook for Digital Signal Processing*, S. K. Mitra and J. F. Kaiser (eds.), John Wiley, New York, N.Y., 1993.

The Cardinal call analyzed in Section 7 is from the audio cassette

> L. Elliot, *Know Your Bird Sounds: Eastern and Central North America*, NatureSound Studio, Ithaca, N.Y., 1991.

For a readable, not-too-technical introduction to the mechanisms of sound production in birds, with plenty of spectrograms similar to the ones in this chapter, see

G. A. Thielcke, *Bird Sounds*, University of Michigan Press, Ann Arbor, Mich., 1976.

Problems

1. Prove that Eq. 2.2, the closed form for a finite geometric series, is correct. (Hint: Use induction.)

2. As pointed out in Section 2, it's obvious that the geometric series in Eq. 2.1 approaches the value n when the ratio between terms approaches one. Prove this another way by applying L'Hôpital's rule to the closed form in Eq. 2.2. If you've never studied the rule, look it up in a first-year calculus book; it's very useful in situations like this, where a zero cancels a pole and we get what looks like 0/0. It was named after the French mathematician Guillaume Francois Antoine Marquis de L'Hôpital (1661–1704).

3. Verify that the frequency content of a finite stretch of a phasor at frequency θ peaks at precisely $\omega = \theta$.

4. Find an expression for the z-transform of a finite stretch of a cosine wave, say $\cos(t\theta)$, $t = 0, \dots, n-1$. Does the frequency content peak at precisely $\omega = \theta$?

5. Show that the Hamming window is normalized to satisfy Eq. 3.6 by dividing by $0.54n - 0.46$.

6. If the 0.54 and 0.46 in the Hamming window are both replaced by 0.5, we get the Hann window:

$$h_t = 0.5(1 - \cos(2\pi t/(n-1))), \qquad 0 \leq t < n$$

Find a closed form for its frequency content and write a program to compute it. Compare the frequency content of the Hann window with that of the Hamming window. Which has the narrower central lobe? What is the relative size of the side lobes?

7. Repeat Problem 6 for the Blackman window:

$$h_t = 0.42 - 0.5\cos(2\pi t(n-1)) + 0.08\cos(4\pi t/(n-1)), \qquad 0 \leq t < n$$

8. Repeat Problem 6 for the Bartlett (triangular) window:

$$h_t = \begin{cases} \dfrac{2t}{(n-1)} & \text{if } 0 \leq t \leq \dfrac{(n-1)}{2} \\[2ex] 2 - \dfrac{2t}{(n-1)} & \dfrac{n-1}{2} < t \leq n-1 \end{cases}$$

9. What is the effect on the frequency content of a signal if every other sample is multiplied by -1? Can you think of some practical use for this simple operation?

10. Offer an explanation for the widening and narrowing of the spectrogram in Fig. 7.7 as time progresses. (This spectrogram used a rectangular window.) Suggest a way to verify your explanation, and try it.

11. Estimate the pitch of the cardinal call before and after the sudden frequency doubling by measuring the periods from Fig. 7.6. Do these check with the spectrograms?

Aliasing and Imaging

1 Taking stock

Moving back and forth between the digital and analog worlds is the trickiest part of signal processing, and the associated effects on signals can be subtle. In this chapter we're going to take a more careful look at aliasing, using the transform methods we've developed since we first introduced the subject in Chapter 3. We're also going to look at digital-to-analog conversion, the process that is the reverse of sampling — and the way in which computers make sound we can hear and pictures we can see.

First, however, I want to step back and take a moment to put all our work in perspective. We've accumulated quite a collection of techniques, and the various domains and transforms might be a little hard for you to keep straight at this point. The fact is, we've constructed the entire foundation of practical signal processing, and this is therefore a good time to review and consolidate. I want you to grasp the main ideas in signal processing as a coherent whole with different incarnations of just a few basic principles. The symmetries and analogies we've been pointing out along the way are a big help.

To a large extent, the art of signal processing is knowing how to move freely and bravely (and correctly) from one domain to another — from continuous to discrete and back, from time to frequency and back. Consider the steps involved in analyzing the cardinal's call in the previous chapter. I started with an analog tape recording, sampled it to obtain a computer file, and then used the FFT for analysis. Figure 1.1 shows all the domains involved in this process, with the time domains in the left column and the frequency domains in the right. The analysis process started at the upper left (the analog time domain), sampled to go down (to the digital time domain), used the z-transform to go right (to the digital frequency domain), and finally went down to get samples of the frequency content (to the DFT frequency domain). As before, circles in this diagram indicate domains that are finite in extent, or equivalently, periodic.

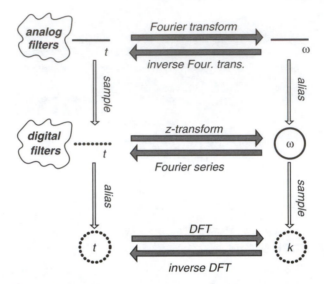

Fig. 1.1 The six domains of signal processing; the time domains are on the left, and the frequency domains on the right. Analog and digital filtering take place at the spots indicated.

This picture includes all the time and frequency domains we're ever likely to need, and suggests the unification I'm after. In every case the forward transform expresses the frequency content of the signal as the projection of the signal onto a basis of phasors. In every case the inverse transform expresses the signal as a sum or integral of the phasors in the basis, weighted by the frequency content.

Throughout this book we've been concentrating on digital processing, and giving the analog domain short shrift. But you needn't feel cheated, because the ideas are interchangeable between domains. For example, up to now, I've alluded only obliquely to the Fourier transform. But you know everything you need to know about it already. The basis signals for the Fourier transform are phasors, of course. Because frequency is a continuous variable, the representation of a signal in terms of the basis of phasors must be an integral and not a sum; because analog frequencies are infinite in extent (there is no sampling and hence no Nyquist frequency) the integral must extend from $-\infty$ to $+\infty$. The representation of an analog signal via the inverse Fourier transform must therefore be

$$x(t) \;=\; \frac{1}{2\pi}\int_{-\infty}^{\infty} X(\omega)\,e^{j\omega t}d\omega \qquad (1.1)$$

It couldn't be anything else. The factor $1/2\pi$ is just a normalizing constant that crops up in inverse transforms. We called it a ''mathematical bad penny'' at the end of Chapter 3, when we peeked ahead at frequency representations. Similarly, the forward

Fourier transform is the projection onto a basis of phasors:

$$X(\omega) \;=\; \int_{-\infty}^{\infty} x(t)\,e^{-j\omega t}dt \tag{1.2}$$

Table 1.1 shows the mathematical form of every transform in Fig. 1.1. As you can see, the formulas are essentially the same, except we need to use sums instead of integrals when we add up discrete items, and opposite signs in the forward and inverse transform phasors.

Transform	Forward	Inverse
Fourier transform	$X(\omega) \;=\; \displaystyle\int_{-\infty}^{\infty} x(t)\,e^{-j\omega t}dt$	$x(t) \;=\; \dfrac{1}{2\pi}\displaystyle\int_{-\infty}^{\infty} X(\omega)\,e^{j\omega t}d\omega$
z-transform	$X(\omega) \;=\; \displaystyle\sum_{k=-\infty}^{\infty} x_k\,e^{-jk\omega}$	$x_k \;=\; \dfrac{1}{2\pi}\displaystyle\int_{-\pi}^{\pi} X(\omega)\,e^{jk\omega}\,d\omega$
DFT	$X_k \;=\; \displaystyle\sum_{n=0}^{N-1} x_n\,e^{-jkn2\pi/N}$	$x_n \;=\; \dfrac{1}{N}\displaystyle\sum_{k=0}^{N-1} X_k\,e^{jkn2\pi/N}$

Table 1.1 The mathematical forms of all the transforms in Fig. 1.1. Allowing for whether the domain is discrete or continuous, finite or infinite, and whether the transform is forward or inverse, these six equations all have the same form. As mentioned in Chapter 9, the z-transform evaluated on the unit circle, shown in the middle row, is sometimes called the Discrete-Time Fourier Transform (DTFT).

There are wonderful symmetries and correspondences between the time and frequency domains, as I've emphasized at several points in preceding chapters. In Fig. 1.1 this means the following:

Any rule stating that an operation in one column corresponds to some operation in the other column, works with left (time) and right (frequency) reversed.

For example, if convolution on the left corresponds to point-by-point multiplication on the right, then convolution on the right will correspond to point-by-point multiplication on the left. We summarize that correspondence thus

$$convolution \;\Longleftrightarrow\; point\text{-}by\text{-}point\ multiplication \tag{1.3}$$

Of course, when convolution is in the time domain, the multiplying function in the frequency domain is called a transfer function, and when convolution is in the frequency domain, the multiplying function in the time domain is called a window — but the operations are mathematically the same.

Furthermore, the rows are analogous in this way:

Any rule that works in any of the three rows also works in the other two.

So, for example, the correspondence in Eq. 1.3 works in all three rows. You can count on these principles every time. Practice using them, and you'll develop intuition that will never let you down.

Thus, we can convolve in any of the six domains, and the convolution is reflected by multiplication of transforms (or inverse transforms) in the opposite domain — the frequency domain if we convolve in the time domain, and the time domain if we convolve in the frequency domain. The term "filtering" is usually reserved for convolution in only two places — the analog or digital time domain, and these operations are called *analog* or *digital filtering*, as indicated in Fig. 1.1. We've already studied these two situations. But we've also seen another example of convolution in the previous chapter: windowing. In particular, we saw that windowing (multiplication) in the discrete time domain corresponds to an appropriate version of convolution in the corresponding frequency domain (see Eq. 6.2 in Chapter 10). I ask you to think about all six kinds of convolution in Problem 3.

2 Time-frequency correspondences

Now I want to be a little philosophical and dig deeper into the real reason why convolution in one domain always does correspond to multiplication in the other. The answer will bring us straight back to Chapter 1 and the properties of phasors.

The way to get more insight into convolution is to consider the effect of a simple time shift. To be concrete, take the case of a digital signal x_k and write it as a "sum" (actually an integral) of phasors, each weighted by the frequency content of the signal at each particular frequency. Mathematically, this is the inverse z-transform, and it appears in Table 1.1:

$$x_k = \frac{1}{2\pi} \int_{-\pi}^{\pi} X(\omega) e^{jk\omega} \, d\omega \qquad (2.1)$$

Now suppose we shift the time variable k by one. The phasor is changed as follows:

$$e^{j(k-1)\omega} = e^{jk\omega} \cdot e^{-j\omega} \qquad (2.2)$$

Shift in time becomes multiplication by a factor that depends on the frequency ω. When the arbitrary signal x_k is shifted, each component in Eq. 2.1 is shifted, and Eq. 2.1 tells us that

$$x_{k-1} = \frac{1}{2\pi} \int_{-\pi}^{\pi} X(\omega) e^{-j\omega} e^{jk\omega} \, d\omega \qquad (2.3)$$

We've seen this many times before: shifting a signal multiplies the transform by a complex exponential. This relation is at the heart of all the important properties of frequency transforms.

Consider convolution again. It is really nothing more than a sum of weighted, shifted versions of a given signal. Therefore, from the relation we've just seen, the result of convolution on the transform is multiplication by a sum of weighted complex exponentials. And that's just the transfer function.

The same thing happens with time and frequency interchanged, because the frequency function can also be expressed as a sum of phasors. In particular, the forward transform analogous to Eq. 2.1 is

$$X(\omega) = \sum_{k=-\infty}^{\infty} x_k e^{-jk\omega} \tag{2.4}$$

which of course is our old friend the z-transform. Shifting the frequency has the effect of multiplying the time function by a complex exponential, and we can just rerun the discussion we've just finished, *mutatis mutandis*.

We have exposed the core of frequency-domain methods. Signals and their transforms can be expressed as sums (or integrals) of phasors, and phasors obey the law

$$shift \iff multiplication\ by\ a\ complex\ exponential \tag{2.5}$$

All the machinery of signal processing that we need in this book relies on this principle. The correspondence between convolution and point-by-point multiplication expressed in Eq. 1.3 is just an elaboration of this more fundamental fact.

Another correspondence that will come in very handy involves even-odd symmetry. Suppose we pick any of the forward or inverse transforms in Table 1.1, and assume that the signal x involved is real-valued, and even about the origin. That is, for every k, $x_k = x_{-k}$, or for every t, $x(t) = x(-t)$. To be concrete, consider the z-transform,

$$X(\omega) = \sum_{k=-\infty}^{\infty} x_k e^{-jk\omega} \tag{2.6}$$

Then the positive and negative terms in the summation can be grouped in pairs, yielding

$$X(\omega) = x_0 + \sum_{k=1}^{\infty} x_k(e^{-jk\omega} + e^{+jk\omega}) = x_0 + \sum_{k=1}^{\infty} 2x_k \cos k\omega \tag{2.7}$$

which is real-valued. We've just proved that if the signal x_k is an even function of time, its transform $X(\omega)$ is real-valued.

The reverse direction is just as simple. Assume that $X(\omega)$ is real-valued, and consider the equation for the inverse transform,

$$x_k = \frac{1}{2\pi} \int_{-\pi}^{\pi} X(\omega) e^{jk\omega}\ d\omega \tag{2.8}$$

Replacing k by $-k$, we get

$$x_{-k} = \frac{1}{2\pi} \int_{-\pi}^{\pi} X(\omega) e^{-jk\omega} \, d\omega \tag{2.9}$$

We're assuming throughout that the signal x_k is real-valued, so taking the complex conjugate of this equation doesn't change the left-hand side. Hence, we're free to take the complex conjugate of the right-hand side, which means simply replacing j by $-j$. That gets us right back to the original right-hand side of Eq. 2.8, so $x_k = x_{-k}$, which means the signal is even. To summarize this argument, a real-valued signal is even if and only if its transform is real-valued.

The preceding proof works for any of the six cases in Table 1.1, and we can express this by another rule:

$$even \iff real\text{-}valued \tag{2.10}$$

We'll be using both this property and the shift-multiply property throughout the rest of the book.

3 Frequency aliasing revisited

Aliasing is an unavoidable issue in digital signal processing, and it can cause problems in unexpected ways. We took a look at the basic idea back in Chapter 3, where we pointed out that sampling means we cannot possibly distinguish between frequencies that differ by multiples of the sampling rate. This simple observation gets at the heart of the matter and explains the need for prefiltering before analog-to-digital conversion. However, we now have the tools to go back and reexamine aliasing from a much more sophisticated perspective. That's definitely worth doing — especially because we also need to understand the process of converting signals from digital back to analog form, where aliasing effects can also cause trouble.

The shift-multiply property provides the key to understanding aliasing in a general setting. The process of sampling an analog signal can be represented as multiplication by a train of ideal pulses, for which we already have a Fourier series, the following sum of phasors (Eq. 6.2 in Chapter 7):

$$b(t) = \frac{1}{T} \sum_{k=-\infty}^{\infty} e^{jk\omega_s t} \tag{3.1}$$

where $\omega_s = 2\pi/T$ is the sampling rate in radians/sec, and the $1/T$ factor normalizes the area of the pulses to unity (the area was T in Chapter 7). Figure 3.1 shows the sampling process from this point of view, multiplication by a pulse train.

When we sample by multiplying in the time domain by the pulse train in Eq. 3.1, each of the component complex exponentials shifts the signal transform. The result in the frequency domain is therefore an infinite sum of shifted versions of the signal's transform. Mathematically, this means that after sampling a signal $x(t)$ its transform becomes

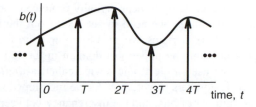

Fig. 3.1 Sampling as multiplication by a pulse train.

$$\frac{1}{T} \sum_{n=-\infty}^{\infty} X(\omega + n\omega_s)$$

(3.2)

where $X(\omega)$ is the transform of $x(t)$. This formula succinctly and elegantly describes the effect of aliasing. To find the new frequency content at a given frequency ω we pile up the frequency content at all frequency points that are displaced from ω by integer multiples of the sampling rate.

Figure 3.2 illustrates aliasing in the frequency domain. It's just like the picture we saw in Chapter 3, of course, but there we reasoned in terms of isolated phasors, and now we can understand the meaning of frequency content in a much more general context.

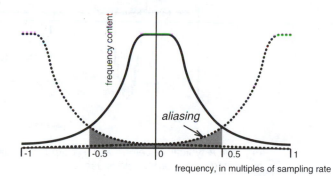

Fig. 3.2 The effects of aliasing in the frequency domain. Versions of the signal spectrum are shifted by integer multiples of the sampling frequency. The versions shifted up and down by the sampling frequency are shown together with the original baseband spectrum. As indicated, frequency components that end up in the baseband from shifted copies of the original spectrum constitute aliasing.

4 Digital-to-analog conversion

If we're ever going to hear digitally produced sound, or, for that matter, see digitally produced images, we need to get back from the numbers to an analog signal. In our picture of six domains, we need to go up from the digital signal domain to the analog signal domain, the uppermost domain in the left column of Fig. 1.1. In the physical world we use an electronic device called a *digital-to-analog (d-to-a) converter* that converts numbers to voltages. As we mentioned in Chapter 3, converters work with a fixed number of bits, and the discrepancy between the theoretically exact signal value and the value represented by that number of bits accounts for a certain amount of quantizing noise. At this point we are more concerned with the ideal operation of conversion and its effect on frequency content.

An interesting question arises immediately: If the digital-to-analog converter produces a voltage at each sampling instant, what voltage should be produced *between* sampling instants? The most common answer is illustrated in Fig. 4.1. The voltage between sampling points is held constant at its most recent value. The circuit that does this is called a *zero-order hold* (because we are interpolating between samples with constants, which are zero-order polynomials). You may also occasionally run into the terms *sample-and-hold* or *boxcar hold*.

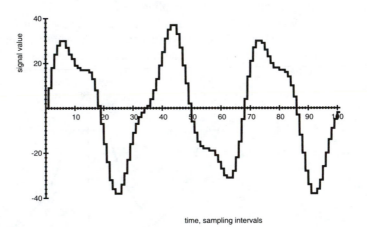

Fig. 4.1 A typical waveform at the output of a digital-to-analog converter, assuming a zero-order hold is used.

A glance at Fig. 4.1 should raise a red flag. It's a very choppy signal, full of discontinuities, and we've learned that jumps produce lots of high frequencies. More precisely, any instantaneous jump in a signal produces components in the spectrum that decay as the reciprocal of the frequency. We might guess, therefore, that the raw output of the digital-to-analog converter sounds bad. It does, and it needs to be processed further. Fortunately, we now understand enough theory to know exactly what to do.

The first step in the analysis of the digital-to-analog converter output is to think of it as the result of applying ideal impulses to an analog filter that has the appropriate impulse response. That impulse response, say $h(t)$, must be the waveform shown in Fig. 4.2 — a constant value for one interval following the impulse, and then zero:

$$h(t) = \begin{cases} 1/T & \text{for } 0 \leq t \leq T \\ 0 & \text{otherwise} \end{cases} \qquad (4.1)$$

It will be convenient to normalize this impulse response so that its area is one, and since its base is T sec wide, we choose its height to be $1/T$, as shown. Figure 4.3 then shows how we think of the operation of the zero-order hold: a digital signal represented by a pulse train, driving a filter that responds to each pulse with a rectangle whose height is proportional to the signal value and whose width is T sec.

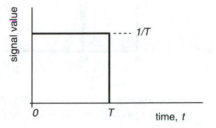

Fig. 4.2 Impulse response of the zero-order hold. The output in response to a unit pulse holds that pulse for exactly one sampling interval of length T. Its height is normalized so that its area is one.

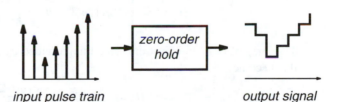

input pulse train   output signal

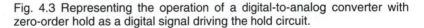

Fig. 4.3 Representing the operation of a digital-to-analog converter with zero-order hold as a digital signal driving the hold circuit.

The rest is easy. We're going to multiply the frequency content of the signal by the frequency response of the zero-order hold. First, we know the spectrum of the digital signal. As we discussed in Section 2 of Chapter 9, the spectrum of the digital signal x at the frequency ω is the inner product of the signal with the basis phasor at that frequency, which gives

$$X(\omega) = \langle x, e^{jk\omega} \rangle = \sum_{k=-\infty}^{\infty} x_k e^{-jk\omega} \qquad (4.2)$$

This, of course, is our old friend the z-transform, evaluated on the unit circle. The frequency content $X(\omega)$ is a periodic function of frequency, again as we discussed in Chapter 9. Its values in the range of frequencies between minus and plus the Nyquist are repeated at all multiples of the sampling frequency, as illustrated in Fig. 4.4.

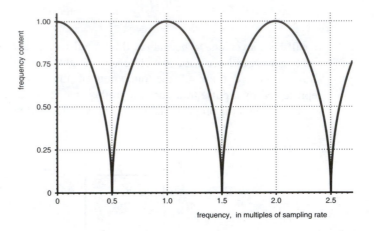

Fig. 4.4 Typical spectrum of a digital signal, the input to the zero-order hold. This spectrum is always periodic, with period equal to the sampling frequency.

The next step is to find the frequency response of the zero-order hold. But that's just the Fourier transform of its impulse response. Using Eq. 4.1, we get

$$H(\omega) = \int_{-\infty}^{\infty} h(t) e^{-j\omega t}\ dt = \int_{0}^{T} \frac{1}{T} e^{-j\omega t}\ dt \qquad (4.3)$$

The integral of $e^{-j\omega t}$ is just the same thing divided by $-j\omega$; after some simplification this becomes

$$H(\omega) = e^{-j\omega T/2}\ \frac{\sin(\omega T/2)}{\omega T/2} \qquad (4.4)$$

It's worth inspecting this result carefully, because it comes up all the time in signal processing, in various forms. First, the factor $e^{-j\omega T/2}$ in front represents a delay of one-half a sampling interval. This delay is a consequence of the fact that the impulse response is centered at the point halfway between $t = 0$ and $t = T$. The remaining factor, $\sin(\omega T/2)/(\omega T/2)$, is a real-valued function of ω, and our discussion at the end of Section 2 showed it must correspond to a signal that is an even function of time. In fact, it corresponds to a rectangular pulse centered at the origin. Finally, Fig. 4.5 shows a plot of the magnitude transfer function of the zero-order hold versus frequency.

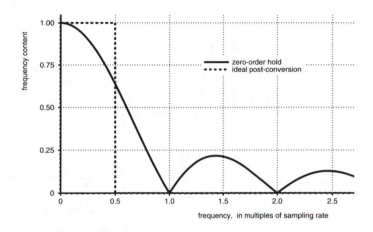

Fig. 4.5 The magnitude transfer function of the zero-order hold. The dashed line shows the ideal post-conversion frequency response.

5 Imaging

In an ideal world, we would choose the frequency response shown as a dashed line in Fig. 4.5, which blocks perfectly all frequencies above the Nyquist and passes perfectly all frequencies below. In contrast, the zero-order hold makes a feeble attempt to remove frequencies above Nyquist, and its magnitude frequency response doesn't actually get down to zero until twice the Nyquist, the sampling frequency. It then bounces back up, and keeps bouncing indefinitely.

The spectrum of the output signal of the zero-order hold, the signal illustrated in Fig. 4.1, is the product of the periodic spectrum of the pulse train, shown in Fig. 4.4, and the transfer function we've just found. The resultant product is shown in Fig. 5.1. This plot verifies what we predicted from the choppy nature of the signal: lots of high frequencies are present, especially in the region just beyond the Nyquist frequency. These components don't belong there — they're called *images* because they are reflections and translations of frequencies in the original signal.

Do we hear imaging? If the sampling frequency is as high as 44.1 kHz, the rate for compact discs, images appear in the region above 22.05 kHz, and probably don't get through most audio systems. In any event, those frequencies are at the very high end of human hearing, and of young humans at that. On the other hand, telephone speech has a much lower bandwidth than CD-quality music, and is sometimes sampled at rates as low as 16 kHz. Imaging just above 8 kHz can be quite objectionable.

If imaging is a problem it can be eliminated with a post-conversion analog filter. However, it is often easier to increase the effective sampling rate digitally, and then simply convert at the higher rate. We'll see how to increase the sampling rate of a digital signal in the final chapter.

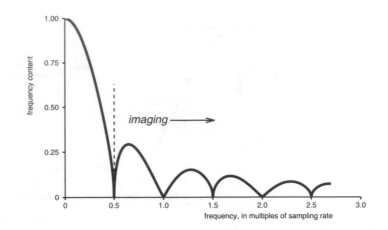

Fig. 5.1 Effect of digital-to-analog conversion on the spectrum. The signal spectrum in Fig. 4.4 is used as an example; the result shown is that spectrum multiplied by the magnitude transfer function in Fig. 4.5. Everything beyond the Nyquist frequency, 0.5 on the abscissa, is imaging.

6 Nyquist's theorem

We now have within easy reach one of the most amazing facts about signals and the way they carry information. It was discovered by Harry Nyquist in 1928, and is why we call the Nyquist frequency the Nyquist frequency.

Aliasing caused by sampling destroys information. The frequencies that are confounded pile up on top of one another, and can never be unraveled. We can't hope to reverse this effect when we go back to the analog domain. The best we can hope for is to do a very good job prefiltering so that we preserve the limited band of frequencies up to the Nyquist — but nothing more. When a signal is perfectly bandlimited, so that it has no frequency components beyond the Nyquist frequency, there is nothing to get confounded, and no information is lost by the sampling process. In that ideal case it is theoretically possible to recover the original signal with absolute perfection. That's the amazing fact, and we're now going to prove it.

Consider the following thought-experiment, illustrated step-by-step in Fig. 6.1. First, imagine that we start with a signal that is perfectly bandlimited, so that it has frequency components only in the range from $-\omega_s/2$ to $+\omega_s/2$, where ω_s is the sampling frequency in radians per sec, as usual. The resultant digital signal has a periodic spectrum, as we've seen many times before. Next, pass it through the ideal lowpass filter with the response shown as a dashed line in Fig. 4.5. This filter gets us back to the original spectrum, and hence back to the original signal. As we mentioned above, there is no aliasing caused by the sampling, and the restoration is perfect. That's the main point of Nyquist's result.

To express this thought-experiment mathematically, we need to derive the impulse response of the ideal lowpass filter. Its frequency response is a constant in the

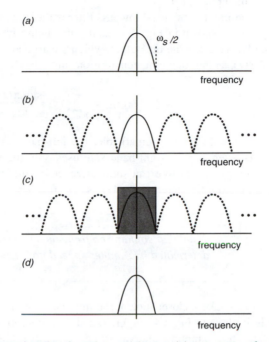

Fig. 6.1 Illustrating the thought-experiment: (a) spectrum of a bandlimited signal; (b) spectrum after sampling; (c) ideal lowpass filtering; (d) back to the original signal.

passband and zero elsewhere:

$$H(\omega) = \begin{cases} T & \text{for } -\omega_s/2 \leq \omega \leq \omega_s/2 \\ 0 & \text{otherwise} \end{cases} \qquad (6.1)$$

This time it's convenient to make the height T, so the area is $T \times \omega_s = 2\pi$. The impulse response $h(t)$ is the inverse Fourier transform of this, which is

$$h(t) = \frac{1}{2\pi} \int_{-\infty}^{\infty} H(\omega) e^{j\omega t} d\omega = \frac{T}{2\pi} \int_{-\pi/T}^{\pi/T} e^{j\omega t} d\omega \qquad (6.2)$$

We encountered this integral in Section 4. It's one of the first you do in first-year calculus, $\int e^{ax} dx = e^{ax}/a$, where in this case the constant $a = jt$, since we're integrating with respect to ω. Evaluating that between the limits of integration gives

$$h(t) = \frac{\sin(\pi t/T)}{\pi t/T} \qquad (6.3)$$

The output of the ideal lowpass filter is the convolution of this, its impulse response, with its input signal.

The input to the ideal lowpass filter is the digital signal consisting of a sequence of pulses, as shown in Fig. 4.3, each pulse being weighted by the signal value x_k. So, finally, we can write the output, which we argued must be identical to the original signal $x(t)$, as the following convolution sum:

$$x(t) \;=\; \sum_{k=-\infty}^{\infty} x_k \, \frac{\sin(\pi(t-k)/T)}{\pi(t-k)/T} \tag{6.4}$$

This formula is truly remarkable: the left-hand side is the signal for any value of t whatsoever, but the right-hand side uses only the values of the signal at the discrete sampling instants. We can summarize the result in terms of the sampling rate f_s Hz and the sampling interval $T = 1/f_s$ sec as follows:

> *A signal that contains no frequencies beyond $f_s/2$ Hz is completely determined by samples spaced no farther apart than $1/f_s$ sec.*

Let's take a closer look at the impulse response of the ideal lowpass filter and the role it plays in Eq. 6.4. Figure 6.2 shows this impulse response plotted versus time t, where the sampling interval is conveniently chosen to be unity, so the sampling instants are the integers. First, we need to settle the question of its value at $t = 0$, when both the numerator and denominator are zero. The answer is one, and it's another application of first-year calculus, using L'Hôpital's rule again (see Problem 9). At sampling instants other than $t = 0$, the $\sin(\pi t/T)$ factor is zero because the argument of the sine is an integer multiple of π and the denominator is not zero. To summarize, the impulse response at the sampling instants kT is

$$h(kT) \;=\; \begin{cases} 1 & \text{if } k = 0 \\ 0 & \text{if } k \neq 0 \end{cases} \tag{6.5}$$

This is a very desirable property for any filter used to reconstruct a continuous signal from samples — at sampling instants it weights the present sample of the signal by one and all other samples by zero. This implies that the output signal will coincide exactly with the original signal at sampling instants.

The astounding part is that the output signal coincides exactly with the original signal at *every* time t, even the times between sampling instants. The convolution sum in Eq. 6.4 tells us exactly how much each of the infinite set of samples x_k must be weighted to determine the value $x(t)$ with absolute precision. This restoration is possible only because the original signal is perfectly bandlimited.

Look at it another way: all the information in a bandlimited signal can be captured by sampling at a frequency equal to twice the highest frequency present in the signal. A bandlimited signal can carry information at that rate, but no higher.

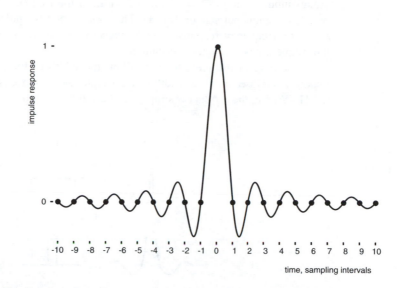

Fig. 6.2 Impulse response of the ideal post-conversion lowpass filter. For this plot the sampling interval is one.

But remember that the ideal lowpass reconstruction filter in Eq. 6.1 is just that: ideal. The frequency response of a real filter can't jump discontinuously, as the response of our ideal filter does, because that would mean passing one frequency and perfectly rejecting another that is infinitesimally close. In practice, as mentioned in the previous section, when imaging is a problem the zero-order hold can be followed by an analog lowpass filter that is an approximation to this ideal.

7 The Uncertainty Principle

In this section we'll indulge in a short digression to examine a particularly pretty example of time-frequency symmetry, and an important aspect of this symmetry. We've seen that the impulse response of the zero-order hold is a rectangular time pulse of width T, and its spectrum is $\sin(\omega T/2)/(\omega T/2)$. We've also seen that the frequency response of the ideal lowpass reconstruction filter is a rectangular frequency pulse of width $2\pi/T$, and its impulse response is $\sin(\pi t/T)/(\pi t/T)$. As far as the shapes of the functions are concerned, time and frequency can be interchanged. In both cases, we have a perfect rectangular pulse in one domain, and what we can think of as an imperfect pulse in the other domain.

What's even more interesting than the symmetry itself is the relation between the widths of these pulses. Let's think of the width of the $\sin x/x$ pulse as the width of its main lobe, which is determined by the value of the first zero-crossing of the sine factor.

In the case of the zero-order hold frequency response, that zero-crossing is the frequency ω at which $\omega T/2 = \pi$, and the width of the frequency response pulse is twice this, $4\pi/T$ radian per sec, or $2/T$ Hz. The shorter the time pulse, the smaller T, and the wider the frequency response. More precisely, the product of the widths of the time and frequency pulses is the constant 2.

In the case of the ideal lowpass filter, the width of the main lobe of its impulse response is $2T$ sec, and the width of the frequency response is $2\pi/T$ radian per sec, or $1/T$ Hz. Again, the product is simply 2. These relationships are illustrated in Fig. 7.1.

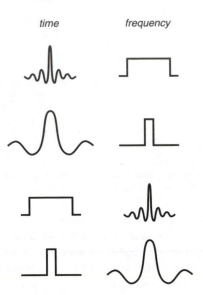

time   *frequency*

Fig. 7.1 Time-frequency symmetry and the Uncertainty Principle — illustrated by the rectangular pulse and its transform. The narrower the pulse in one domain, the wider in the other. A rectangular pulse in one domain corresponds to a $\sin(x)/x$ shape in the other. The first two rows correspond to ideal lowpass filtering, the last two to the impulse response and frequency response of a zero-order hold.

This is an instance of a principle that permeates many areas of science, and is often referred to as the *Uncertainty Principle*. In signal processing, it means that the narrower a time pulse, the wider its frequency content, and vice versa. In quantum mechanics, where the principle was first enunciated by the German physicist Werner Heisenberg (1901–1976), it means, for one thing, that the more precisely we measure the position of a particle, the less certain we can be about its momentum, and vice versa. The mathematics behind both results is the same.

We saw another example of the Uncertainty Principle when we discussed windowing for the FFT. The wider the window in the time domain, the narrower the averaging in the frequency domain, and hence the better the frequency resolution. That's why the 200-inch telescope at Mount Palomar has better resolution than your 6-inch telescope.

Yet another example of the Uncertainty Principle is the two-pole reson filter discussed in Chapter 5. The narrower its bandwidth, the more slowly its impulse response decays.

8 Oversampling

You might think the only way to avoid harmful aliasing when doing analog-to-digital conversion is to build a very good analog lowpass prefilter. That's usually an expensive and troublesome proposition, and fortunately there's a profitable way to trade off analog filtering before sampling with digital filtering after sampling. The method is usually referred to as *oversampling*, and gives us a good opportunity to learn more about aliasing. It will also make you want to read the next two chapters so that you can design good digital filters.

To take a particular example, suppose we want to convert an analog signal to digital form with the ultimate sampling rate of 40 kHz. We already know that frequency content above the Nyquist frequency of 20 kHz will be folded down into the baseband, as illustrated once more in Fig. 8.1(a). The new idea is to convert at a higher rate to avoid aliasing, and then use digital processing to get a cleaner digital signal at the final desired sampling rate. The extra digital filtering will almost invariably be less expensive than the good analog prefiltering required at the original rate. This whole scheme is cost-effective if the faster analog-to-digital converter is not too expensive, which is usually true, at least at audio frequencies.

Figure 8.1(b) shows the spectrum if we sample at 80 kHz, twice the final desired rate. The Nyquist frequency at this rate is 40 kHz, and it's usually trivial to knock out any frequency components above that. In fact, most audio systems will do that without your worrying about it. So far, we have avoided aliasing.

The next step is to trim the spectrum of the resultant signal with a digital filter. We want to eliminate any frequency content above the final Nyquist frequency of 20 kHz. The desired digital filter response is indicated in Fig. 8.1(c) by shaded area.

We now have an interesting situation. The signal fills only half its allotted bandwidth. It is redundant in the sense that it is represented by twice as many samples as required by Nyquist's theorem. The next step is surprisingly easy: We simply throw away every other sample. To see what effect this has on the signal's frequency content, look at the spectrum $X^{\#\#}$ that results when a signal with spectrum X is sampled at frequency $2\omega_s$:

$$X^{\#\#}(\omega) = \frac{2}{T}\left[\cdots + X(\omega-2\omega_s) + X(\omega) + X(\omega+2\omega_s) + \cdots\right] \quad (8.1)$$

(I'll use this weird notation only in this section.) This is just the aliasing formula, Eq. 3.2, with sampling frequency $2\omega_s$ and sampling interval $T/2$. Compare this with the spectrum that results from sampling at rate ω_s:

$$X^{\#}(\omega) = \frac{1}{T}\left[\cdots + X(\omega-\omega_s) + X(\omega) + X(\omega+\omega_s) + \cdots\right] \quad (8.2)$$

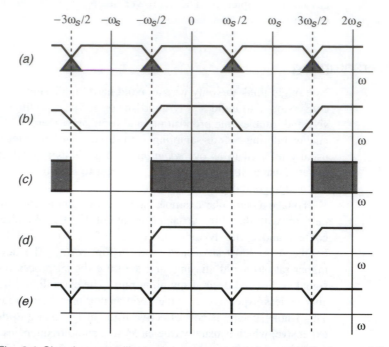

Fig. 8.1 Signal spectra illustrating oversampling to avoid aliasing, and the digital processing after sampling to reduce the sampling rate: (a) sampling at the rate ω_s results in aliasing; (b) sampling at $2\omega_s$ doesn't; (c) the desired digital filter frequency response to prepare for sampling rate reduction; (d) the signal spectrum after digital filtering; (e) the final spectrum after discarding alternate samples, showing that aliasing has been avoided.

Discarding alternate samples converts the signal's spectrum from $X^{\#\#}$ to $X^\#$. This is because throwing away every other sample gets us to exactly the same signal as sampling at half the rate to begin with.

The relationship between the spectra in Eqs. 8.1 and 8.2 is actually very simple. Except for a factor of 2, the spectrum $X^{\#\#}$ consists of every other term of the spectrum of $X^\#$. We can therefore get $X^\#$ by adding $X^{\#\#}$ to a version of $X^{\#\#}$ shifted by ω_s, to fill in every other image. This interleaves copies of the spectrum at the correct spacing of ω_s. Therefore,

$$X^\#(\omega) \;=\; \frac{1}{2}\Big[\, X^{\#\#}(\omega) \;+\; X^{\#\#}(\omega-\omega_s)\,\Big] \tag{8.3}$$

You can think of this formula as expressing an operation of *subaliasing*. Striking out every other sample is a very mild version of the sampling that gets us from analog to digital signals. Instead of producing an infinite number of images of the spectrum, it produces only two. The effect of this process of weeding out alternate samples is illustrated in Fig. 8.2, and occurs in the transition from Fig. 8.1(d), which shows $X^{\#\#}$, to Fig. 8.1(e), which shows $X^\#$.

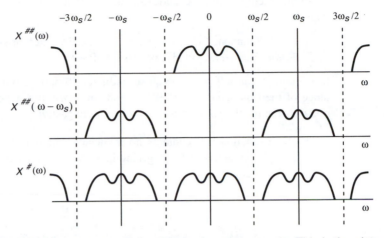

Fig. 8.2 The effect of throwing away every other sample. This is the picture corresponding to Eq. 8.3, and explains how we get from Fig. 8.1(d) to Fig. 8.1(e). The first two spectra are added to produce the third, an effect we can call subaliasing.

What we need to make oversampling work is the digital filter with the frequency response shown shaded in Fig. 8.1(c). At that point the sampling frequency is $2\omega_s$, the Nyquist frequency is therefore ω_s, and the lowpass digital filter is designed to pass frequencies up to only half this Nyquist frequency. Designing filters like this is the subject of the next two chapters.

Notes

Nyquist saw through to the fundamental connection between the rate at which we can send information and the bandwidth of the channel in the famous paper

H. Nyquist, "Certain Topics in Telegraph Transmission Theory," *Trans. Amer. Inst. of Elect. Eng.*, vol. 47, pp. 617–644, April 1928.

He puts things in terms of telegraph waves and bases his main argument on Fourier series instead of on the more general Fourier transform, but the main point is unmistakable. Nyquist puts it this way:

"The minimum band width required for unambiguous interpretation is substantially equal, numerically, to the speed of signaling...."

In our terms, the "band width" stretches from $-\omega_s/2$ to $\omega_s/2$, and is therefore equal to ω_s, the sampling rate, and his "speed of signaling." It's important to realize that this band of width ω_s doesn't need to be centered around zero frequency. In radio transmission, for example, it's centered around the carrier frequency of the station. But the rate at which information is transmitted is still determined by the bandwidth of the signal.

What we called Nyquist's theorem in this chapter is sometimes called the Shannon sampling theorem, after Claude Shannon (1916–), because of the equally famous paper

> C. E. Shannon, "Communication in the Presence of Noise," *Proc. Inst. Radio Engineers*, vol. 37, pp. 10–21, 1949.

But Shannon himself gives Nyquist credit for "pointing out the fundamental importance of the time interval $1/(2W)$ seconds in connection with telegraphy...." Shannon puts the result in a form we recognize more easily:

> "If a function $f(t)$ contains no frequencies higher than W cps [Hz], it is completely determined by giving its ordinates at a series of points spaced $1/2W$ seconds apart."

Problems

1. Prove that it doesn't matter whether we window a continuous-time signal and then sample, or sample and then window.

2. Figure 1.1, which shows the six usual domains of signal processing, really omits two, as you will see if you refer to the beginning of Chapter 9. What two domains have we omitted, and why are they less important than the other two for the usual kinds of signals processing?

3. Write out the mathematical expression for convolution in each of the six domains shown in Fig. 1.1. Then write explicitly the equivalent operation in the corresponding transform domain, the horizontal partner in that figure.

4. The key property for basis functions for frequency transforms, as argued in Section 2, is the following. Let $U(k) = e^{j\omega k}$, thinking of ω as a constant for this problem, and let k be the discrete time variable, as usual. Then the key property in Eq. 2.2 takes the form

$$U(k-1) = U(-1)U(k)$$

Prove that the only function that satisfies this relation for all k is the exponential function of the form

$$U(k) = c^k$$

where c is some constant.

5. The opposite of an even function is an odd function, by which we mean that $x_k = -x_{-k}$. The opposite of real-valued is imaginary-valued. Demonstrate that the following time/frequency correspondence rule is valid:

$$odd \iff imaginary\text{-}valued$$

6. Go back to the picture of the six domains in Fig. 1.1, and notice that the sampling of the spectrum implied by the DFT calculation is reflected by an aliasing operation in the time domain, in keeping with our correspondence principles.

It works this way: sampling in the frequency domain means we consider only the set of frequencies that are integer multiples of some fixed f_0 Hz. The resulting time waveform is a Fourier series, and is periodic with period $1/f_0$ sec. We have therefore confounded signal values that are spaced $1/f_0$ sec apart in time. The effect is perfectly analogous to sampling in the time domain and confounding frequencies.

Derive a mathematical expression for this time aliasing.

7. Explain why there is no time aliasing in the usual applications of the DFT calculation.

8. In some mathematical contexts the aliasing operation on the transform represented by Eq. 3.2 is called the *cylinder* operation. Why?

9. Use L'Hôpital's rule to check that the impulse response of the ideal post-conversion filter is one at $t = 0$.

10. Rather than keep the output of a digital-to-analog converter constant between samples, we might connect adjacent sample values with a straight line between them.

(a) Show that the impulse response of such a post-conversion "hold circuit" is an isosceles triangle with base extending from time instants $-T$ to $+T$.

(b) Show that the convolution of a rectangular pulse with itself is also an isosceles triangle.

(c) Combine the results of Parts (a) and (b) to find the frequency response of this particular hold circuit.

11. You can think of aliasing as the confounding of frequencies. The process of analog-to-digital conversion confounds a given frequency with every frequency that differs from it by an integer multiple of the sampling frequency. What frequencies are confounded with a given frequency when alternate samples are thrown away?

12. Eq. 8.3 tells us the effect of subsampling by using every second sample value. Generalize it to the scheme that uses only every kth sample, where k is an integer that can be larger than 2. What is the cutoff frequency of the lowpass digital filter we need in the corresponding oversampling scheme?

CHAPTER 12

Designing Feedforward Filters

1 Taxonomy

Practical filter design is a highly refined and continually evolving combination of science and art. Its roots go back to the development of analog filters for radio and telephone applications in the 1920s. Thousands of papers and dozens of books have been devoted to the subject, and effective programs are available that implement the important algorithms. The subject has its own fascination, and connoisseurs spend rainy Saturday afternoons playing with exotic filter design algorithms. But if you stick to the common applications of digital signal processing, 99 percent of the filters you'll ever need can be designed beautifully using just two basic approaches, one for feedforward filters and one for feedback filters.

Just a word about what I mean by "design." In a sense we've already designed some digital filters — two-pole resonators, plucked-string filters, and allpass filters, for example. But the term "digital filter design" has come to mean a certain general approach to the design process. Most digital signal processing practitioners take it to mean design from stipulated specifications using a well-defined optimization criterion. Most often, the specification is in terms of the magnitude frequency response, and the optimization criterion is to find the filter whose response deviates least from the desired specification. By contrast, our previous designs can be considered only ad hoc.

We've just seen one example of how we might specify a digital filter by its desired frequency response: the ideal lowpass filter in Fig. 8.1(c) in Chapter 11, which was used as a prefilter before reducing the sampling rate. That ideal response cannot be attained with an actual filter, because it jumps discontinuously, so it must be approximated. And this approximation process is the interesting part of the filter design problem.

A filter of any kind has a certain cost associated with its implementation. Usually, the main cost of using a digital filter can be measured by the time it takes to find each of its output samples, and this is reflected well in the number and type of arithmetic operations required. But there are other costs. Some filters require more precision in the arithmetic operations themselves, for example, and sometimes that is an important determining factor — especially if we plan to use special hardware for the filter implementation.

The filter problem then comes down to balancing the accuracy of the approximation against the cost of actually using the filter. It's a classic example of what is called an *optimization problem*.

I now want to make a distinction between two ways of solving this problem. The first is to be clever and find the solution in some compact, convenient form. For example, for some particular design problem we might be able to find out exactly where to put the zeros of a feedforward filter so that the resulting frequency response is absolutely the best possible, according to some given criterion of closeness to the specifications. We'll have to prove mathematically that this is so, of course. I'll call this a *closed-form* solution to a filter design problem.

The other possible way of solving the problem is to use some kind of iterative technique that keeps improving some initial solution, which may not be very good, until it can't be improved any more. I'll call that an *iterative* solution. Usually, we resort to an iterative solution only when we're convinced that finding a closed-form solution is hopeless. This is the normal scenario in many areas where design problems come up: we try to figure out the answer, and then resort to iterative numerical methods when we give up.

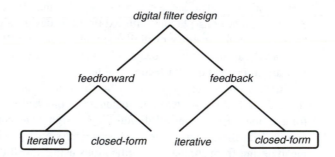

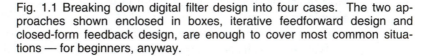

Fig. 1.1 Breaking down digital filter design into four cases. The two approaches shown enclosed in boxes, iterative feedforward design and closed-form feedback design, are enough to cover most common situations — for beginners, anyway.

We can now categorize filter design problems in two ways, depending on whether we want feedforward or feedback filters, and on whether we use an iterative numerical method or seek a closed-form solution. That makes four possibilities, which are illustrated in Fig. 1.1. As mentioned above, it turns out that two of the four cases are big successes, and the two resulting methods provide enough horsepower to cover your basic needs for quite a while: they are iterative design for feedforward filters and closed-form design for feedback filters. I gave examples of the fruits of these design algorithms back in Section 4 of Chapter 4 and Section 8 of Chapter 5, respectively, just to make sure you realized that it's possible to design large and powerful filters of both the feedforward and feedback types. In this chapter we're going to go more deeply into the methodology of designing feedforward filters.

2 The form of feedforward filters

Design problems like the ones we're considering break down into two stages. First, we need to choose the type of filter, and second, we need to choose particular values for its coefficients. In this section, we'll discuss the form we'll be using for feedforward filters. We'll discuss their design in the next section.

Remember that a feedforward digital filter is specified by an update equation of the form

$$y_t = a_0 + a_1 x_{t-1} + a_2 x_{t-2} + \cdots + a_{n-1} x_{t-(n-1)} \qquad (2.1)$$

giving the output sample y_t at time t in terms of the input samples x_t. The filter has n coefficients, and we'll call n the filter *length*. Terminology varies in this regard; it's a matter of taste. Just remember that we're counting from 0, so the last coefficient of a length-n filter is a_{n-1}. The transfer function is

$$\mathcal{H}(z) = a_0 + a_1 z^{-1} + a_2 z^{-2} + \cdots + a_{n-1} z^{-(n-1)} \qquad (2.2)$$

We're now going to make an assumption that will greatly simplify the design job. The idea is based on the property of transforms derived in Section 2 of Chapter 11:

$$even \iff real\text{-}valued \qquad (2.3)$$

(I told you it would come in handy.) We can't quite make the coefficients even, because that means $a_i = a_{-i}$, and the indices start from 0. But we can do something just as good: we'll make the coefficients symmetric about their center. As we'll see shortly, that will make the frequency response real except for a linear-phase factor that represents a delay. We can then forget about the delay factor and concentrate on the rest of the transfer function, which will be real-valued.

To make it easy to see what's going on, let's consider the specific case of a length-5 filter, when the transfer function is

$$\mathcal{H}(z) = a_0 + a_1 z^{-1} + a_2 z^{-2} + a_3 z^{-3} + a_4 z^{-4} \qquad (2.4)$$

We've been in similar situations before, and the standard trick is to factor out a power of z corresponding to the average of the first and last exponents, the average delay, in this case z^{-2}. (This should be second nature by now. We've already seen this maneuver in Section 7 of Chapter 4, Sections 2 and 5 of Chapter 10, and Section 4 of Chapter 11.) The resulting rearrangement is the following completely equivalent transfer function:

$$\mathcal{H}(z) = z^{-2} \left[a_0 z^2 + a_1 z^1 + a_2 + a_3 z^{-1} + a_4 z^{-2} \right] \qquad (2.5)$$

The corresponding frequency response is obtained, as usual, by setting $z = e^{j\omega}$:

$$H(\omega) = e^{-2j\omega}\left[a_0 e^{2j\omega} + a_1 e^{j\omega} + a_2 + a_3 e^{-j\omega} + a_4 e^{-2j\omega}\right] \qquad (2.6)$$

If we assume symmetry of the coefficients, $a_1 = a_3$, so the second and fourth terms combine to form $2a_1\cos\omega$; and $a_0 = a_4$, so the first and fifth terms combine to form $2a_0\cos(2\omega)$. This yields

$$H(\omega) = e^{-2j\omega}\left[a_2 + 2a_1\cos\omega + 2a_0\cos(2\omega)\right] \qquad (2.7)$$

The important point is that the factor inside the parentheses is real-valued, and the factor in front is a complex exponential that represents nothing more than a delay of two samples.

The process of factoring out the delay of two samples has a simple interpretation in terms of the transfer function. Figure 2.1 shows the flowgraph for the original filter, Eq. 2.4. The delayed versions of the input signal are fed forward, true to the name "feedforward." Figure 2.2 shows (to the right of the dashed line) the flowgraph corresponding to $z^2 \mathcal{H}(z)$, the transfer function inside the parentheses in Eq. 2.5. The filter has a "future" term for each past term, and that makes the coefficients even and the transfer function real-valued. Of course the input is delayed two samples to begin with, so the filter does not need to be clairvoyant. The future terms are just in the future with respect to the center term.

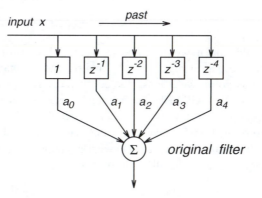
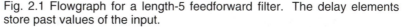

Fig. 2.1 Flowgraph for a length-5 feedforward filter. The delay elements store past values of the input.

As far as the magnitude of the frequency response in Eq. 2.7 is concerned, the complex factor representing the delay is immaterial — it has magnitude one. The only thing that matters is the cosine series inside the parentheses. To simplify the notation still further, we'll use coefficients c_i that run in reverse order, so we can rewrite Eq. 2.7 as

$$\hat{H}(\omega) = e^{2j\omega}H(\omega) = c_0 + c_1\cos\omega + c_2\cos(2\omega) \qquad (2.8)$$

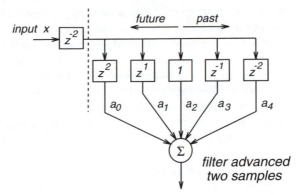

Fig. 2.2 Filter equivalent to the one in the previous figure. A delay of two samples has been inserted at the input, and the filter to the right of the dashed line can be regarded as using future inputs. When the coefficients are symmetric about the center, the frequency response of the advanced filter is real-valued.

where $c_0 = a_2$, $c_1 = 2a_1$, and $c_2 = 2a_0$, and we've moved the pure delay factor to the left-hand side of the equation (where it actually represents an advance, not a delay). The new frequency response $\hat{H}(\omega)$ incorporates this time shift, and is real-valued. We'll use $\hat{H}$ when we put constraints on the frequency response in the next section.

To make life even simpler, we're going to assume that the filter length n is always an odd integer. The case for n even is very similar, and adds no new ideas. With that, it's easy to see that the general form of Eq. 2.8 for a filter of odd length n is

$$\hat{H}(\omega) = e^{jm\omega} H(\omega) = c_0 + c_1 \cos\omega + c_2 \cos(2\omega) + \cdots + c_m \cos(m\omega) \qquad (2.9)$$

where $m = \frac{1}{2}(n-1)$, the number of terms to one side of the center term. Since we count from 0, there are $m+1 = \frac{1}{2}(n+1)$ coefficients c_i. Because we assume symmetry, that's how many coefficients we are free to adjust to achieve a given desired frequency response.

One final point. When we consider the frequency response $H(\omega)$, we should really use the magnitude. But we'll usually use the part without delay, the right-hand side of Eq. 2.9, which is real-valued, and can be negative as well as positive. When it's negative, we have a perfect right to consider it a positive magnitude with a phase angle of π radians, but we won't do that. What's important is that it's real.

To summarize: we're assuming that the feedforward filters have an odd number of terms (n), and have coefficients that are symmetric about their central term. The frequency response is then determined by the real-valued cosine series in Eq. 2.9, with $m = \frac{1}{2}(n+1)$ unknown coefficients c_i. The design problem then boils down to choosing those coefficients to satisfy given specifications — which is next on the agenda.

3 Specifications

We're now going to think about designing a digital filter we have some use for — the one that is used after oversampling in Section 8 of Chapter 11. The purpose of this filter is to eliminate the frequency components in the range of frequencies from one-half the Nyquist frequency to the Nyquist frequency, in order to avoid subaliasing when we drop every other sample to halve the sampling rate. We'll call this a *half-band* filter. The filter's ideal frequency response is just one in the first half of the baseband, and zero in the second half. You know very well, however, that it isn't possible to achieve this ideal with an actual filter. The problem we're considering here is how to select a filter length and a set of coefficients for a feedforward filter to do the job well enough.

No matter how we implement the filter, the more coefficients there are, the more multiplications we're going to need to get each output sample. So the problem comes down to this:

> *Given a precise statement of what frequency response we consider acceptable, find the filter with the fewest coefficients that meets those requirements.*

Specifying what we consider acceptable couldn't be simpler. We just decide on bands of frequency — like passbands and stopbands, and stipulate that the frequency response lie within certain limits in those bands. For example, Fig. 3.1 shows the specifications for the half-band filter we're thinking about designing. The limits are represented by barriers, shown shaded in the figure. We'll use $\hat{H}$ to denote the real-valued frequency response of the centered filter, as in Eq. 2.8. The two barriers in the passband stipulate that $\hat{H} \leq 1.05$ and $\hat{H} \geq 0.95$. Similarly, the stopband requirements are $\hat{H} \leq 0.05$ and $\hat{H} \geq -0.05$. This specification in the stopband, by the way, is a good example of how the frequency response is allowed to be negative or positive.

The next step in our example is actually finding the shortest-length filter that satisfies the constraints in Fig. 3.1. Before I describe how that's done, I want you to stop and think about the consequences of choosing specifications.

Consider the choice of passbands and stopbands. In Fig. 3.1, the passband extends to only 0.23 times the sampling frequency, and the stopband picks up from 0.27 times the sampling frequency, after a gap called a *transition band*. Why be so timid? Why not jam the stopband right up against the passband? The answer lies in the fact that the frequency response is only the beginning of a Fourier series, as you can see from Eq. 2.9. We've already seen in Chapter 7 what happens when we add up a finite number of terms in a Fourier series. Recall the approximations to a square wave. The jumps between zero and one become steeper and steeper as the number of terms increases, but can never be perfectly discontinuous. The closer the upper edge of the passband is to the lower edge of the stopband, the steeper the jump, and the more terms we need in the filter. This makes economic sense: the better the performance in distinguishing between frequencies that are close together, the more expensive the filter.

The same reasoning applies to the ripples. The closer we want the frequency response to the ideal flat response, the more terms we are going to need in the filter.

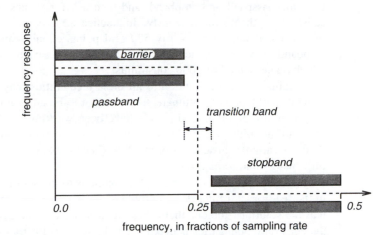

Fig. 3.1 Specification of a half-band lowpass filter, illustrating the use of barriers. The dashed lines show the ideal response, and we want to find the filter that stays inside the barriers and has the fewest coefficients. The passband in this case is [0.0, 0.23] and the stopband is [0.27, 0.5]; the barriers are positioned in those bands at distances ±0.05 from the ideal response.

When we see the results of an actual design algorithm, we'll be able to get a feeling for the cost of good performance.

4 A design algorithm: METEOR

As I stated at the beginning of the chapter, the general design problem for feedforward filters is solved, at least for the situations you're likely to encounter in day-to-day signal processing. The approach I'm going to describe uses a mathematical formulation called *linear programming*, and in particular the program METEOR, as mentioned in Chapter 4. Other iterative numerical methods are faster, but linear programming is the most flexible method available, and is perfectly suited to our formulation of the design problem.

The key to the solution method is the fact that the frequency response, the cosine series in Eq. 2.9, is a *linear* function of the unknown coefficients. Let's take a very simple example, just to illustrate the idea. Suppose we want to design a filter of length 3, which will have only two coefficients for us to choose. The constraint that the frequency response be less than 1.05 in the passband is

$$\hat{H}(\omega) = c_0 + c_1 \cos\omega \leq 1.05 \tag{4.1}$$

For any particular fixed value of frequency ω, this is a linear function of the unknown coefficients c_0 and c_1. Lower bounds are of the same form; for example, in the passband we'll have constraints like

$$\hat{H}(\omega) = c_0 + c_1 \cos\omega \geq 0.95 \tag{4.2}$$

We throw together lots of inequality constraints like these, for lots of values of ω in both the passband and stopband, and then ask for values of the coefficients c_i that satisfy all of them simultaneously. In practice we may want to use a grid of frequency points that is quite fine — say 500 grid points equally spaced in the passband and stopband of our half-band filter. There are two constraints for each frequency point, which makes a total of 1000 constraints.

Finding a feasible solution to all these inequalities may sound like an incredibly difficult problem, but, fortunately, it turns out to be an example of the classical linear programming problem, and people have been working on it since the 1940s. Today there are algorithms that will solve problems with thousands of constraints and hundreds of variables in seconds or minutes. (See the Notes at the end of the chapter for a little history and some references.)

Not only is it easy to find feasible solutions to the set of inequalities that comes up in the feedforward filter design problem, but it's also easy to find, from among all the feasible solutions, the one that is best in the sense that the closest distance from any of the constraint boundaries is maximized. Figure 4.1 illustrates the idea. A collection of inequalities is shown in two dimensions, corresponding to a filter with two coefficients. Any point in the region interior to all the constraints represents a feasible filter design. A point whose closest constraint is as far as possible is also shown — that choice of filter coefficients has a frequency response that stays the farthest away from the boundaries chosen for this problem. The figure is drawn in two dimensions to be easy to grasp. In practice, a typical filter might have, say, a length of 31, and therefore 16 coefficients, so the feasible region would be a polytope in 16-dimensional space. Such a problem is impossible to visualize, of course, but linear programming algorithms have no problem dealing with it.

Before we look at an example, I need to be a little more precise about what I mean by ''distance'' from a constraint. We're not really interested directly in ordinary, Euclidean distance in the space of coefficients, distance in the plane shown in Fig. 4.1, for example. What we're really concerned about is the difference between the filter frequency response and the specification, the difference between the left- and right-hand sides in Eqs. 4.1 and 4.2. We can put the problem in precise mathematical form by inserting a ''squeezing'' parameter δ in those inequalities as follows:

$$\hat{H}(\omega) = c_0 + c_1 \cos\omega \leq 1.05 - \delta \qquad (4.3)$$

$$\hat{H}(\omega) = c_0 + c_1 \cos\omega \geq 0.95 + \delta \qquad (4.4)$$

Remember that we have a pair of inequalities like this for every frequency point on some grid, in both the passband and the stopband. The variable δ represents the true distance between each of those constraints and the frequency response at that point. We therefore try to maximize δ subject to all these constraints.

This now is the precise form of linear programming optimization problem we're interested in:

Find, from among all the feasible sets of coefficients, one that maximizes the minimum distance from the frequency response to any of the constraints.

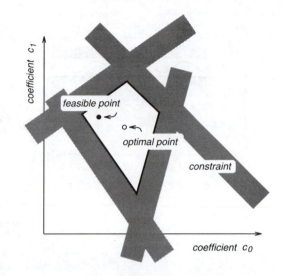

Fig. 4.1 This is what the feedforward filter design problem looks like in the coefficient plane, for two coefficients c_0 and c_1. There is one constraint for each frequency point on a grid of frequency points. The set of feasible coefficients is the region interior to the constraints. The solid dot is a feasible solution, and the open dot is the best feasible solution in the sense that its closest distance to a constraint is maximum.

As promised, there are good algorithms for solving this problem with hundreds of coefficients and thousands of constraints generated by a frequency grid in all the passbands and stopbands. Sometimes this form of optimization problem is called a *minimax* — or, in this case, a *maximin* — problem, because we want to maximize the minimum distance to a constraint.

5 Half-band example

METEOR tries different filter lengths to determine the shortest one that satisfies the constraints. The program gives us a choice between odd and even lengths, and in this case I asked for an odd length. Figure 5.1 shows the frequency response of the answer, given the design specifications shown in Fig. 3.1. The smallest odd-length filter that meets the specifications turns out to be of length 31, which means there are 16 free coefficients.

You will often see the frequency response of filters drawn with a dB scale, and I've redrawn Fig. 5.1 that way in Fig 5.2. It looks different, but it's precisely the same response. The compressed scale makes the passband ripples look smaller than the stopband ripples, even though they are the same size arithmetically. The dB scale also takes the magnitude of the response, so the distinction between the positive and

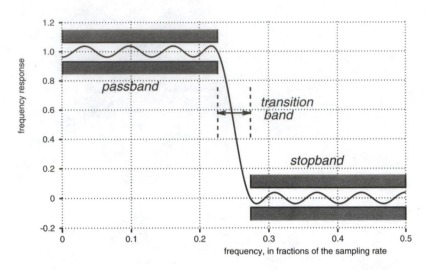

Fig. 5.1 Frequency response of a half-band feedforward filter. The passband is [0.0, 0.23], the stopband [0.27, 0.5], and the required maximum deviations are ±0.05 in both bands. The minimum number of coefficients that meets the specification is length $n = 31$.

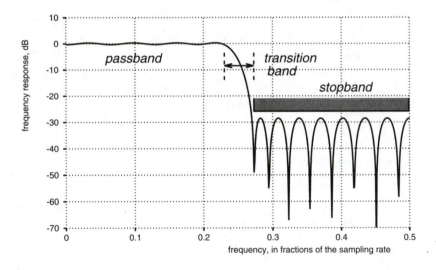

Fig. 5.2 Exactly the same frequency response as the previous figure, except the ordinate is in dB.

negative ripples in the stopband is lost. Sometimes passband ripple is specified in dB, although I find that a bit obfuscating. In particular design examples here I'll stick to arithmetic specification in passbands, like 1 ± 0.05 in this example.

While we're on the subject of dB scales, I should mention that stopband ripple measured in dB is often referred to as stopband *rejection*. For example, if the ripple in a stopband is specified to be at most ± 0.01, we say we require at least 40 dB rejection in that band.

It's worth scrutinizing the optimal response closely, because it is typical of filters that are actually used in practice.

The most striking feature is that the response ripples regularly between upper and lower bounds. In fact, the distances between the tops and bottoms of the ripples and to the constraint boundaries are all equal. This property is called *equiripple*. The following heuristic argument explains why optimal solutions are equiripple.

Suppose the ripples are not equally distant from the constraints, but that one of them, say the *i*th, is closer than all the rest (see Fig. 5.3). We should then be able to adjust the coefficients so that this *i*th ripple moves away from the boundary. (Just jiggle the coefficients. If it moves closer, jiggle them the other way.) We can keep doing that until the *i*th ripple becomes tied with some other ripple for being closest to a boundary.

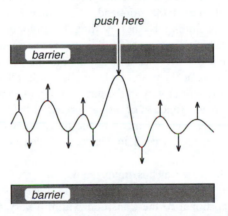

Fig. 5.3 Illustrating the argument that the optimal solution is equiripple. Push on the worst ripple and it becomes better at the expense of the others.

We can repeat the process, but when we jiggle the coefficients, we need to do it in such a way that the two ripples now tied for minimum distance to a boundary stay tied. This will be possible if there are enough coefficients. Eventually, there will be so many ripples tied for closeness to a boundary that we can no longer jiggle the coefficients, and that's when we've reached optimality. (The same kind of argument can be made by looking at the geometry in Fig. 4.1. See Problem 5.)

I want to make it clear that this is more of a plausibility argument than an actual proof, but it offers some insight into why the solutions look the way they do. For one thing, it suggests that the number of ripples tied for closeness to a constraint in an optimal solution is very closely related to the total number of coefficients we have to play with in the filter, the number of terms in the cosine series Eq. 2.9, $m = \frac{1}{2}(n+1)$. In fact, the number of ripples is always one or two more than m (see Problem 7 and the Notes).

6 Trade-offs

The kind of filter we've just designed — with a single transition between two bands — is very useful, and this simple design situation comes up all the time. The principal measure of how much it costs to use a particular feedforward filter is the number of terms, so it's nice to have a rough idea of the filter length that will be required to meet some given specifications. Such estimates will also help us understand the extent to which narrow transition widths and small specified ripples in passbands and stopbands require filters with long lengths.

There's no way known to predict the optimal filter length exactly, without actually doing the design, but there are very good approximations based on empirical experimentation. For example, let's see how the filter length varies with the transition width, when all the other specifications are kept the same as the ones in the lowpass example in Section 5. That is, we'll keep the passband and stopband ripples fixed at ±0.05, the upper end of the passband fixed at 0.23 times the sampling frequency, and vary the lower end of the stopband. We'll then run METEOR for a sequence of transition widths, finding the shortest filter length for each case.

What do we expect to happen to the required filter length as the transition width decreases? Well, certainly it should increase, but at what rate? Let's plot the points and look at them — maybe we'll get an idea from the picture. Since we expect the filter length to vary in the direction opposite that of the transition width, we'll plot the length as a function of the *reciprocal* of the transition width, as shown in Fig. 6.1. (By the way, since we measure the transition width in fractions of the sampling rate, its inverse, the abscissa in Fig. 6.1, is measured in units of multiples of the sampling period.)

A glance at this figure shows that the empirical relationship is quite close to a straight line through the origin. In other words, the required filter length is inversely proportional to the transition width, a very useful and intuitively appealing rule of thumb.

We can relate this result to our earlier discussion of resolution. The transition width is the gap between the frequencies that are passed by the filter and those that are rejected. This gap can be thought of as the *resolution* of the filter in the sense that the filter is able to separate frequencies this close, but no closer. Thus, our empirical result is that the resolving power of a feedforward filter is inversely proportional to its length — just as the resolving power of a telescope mirror is inversely proportional to its diameter (see Section 3 of Chapter 10 and the Notes for that chapter).

The constant of proportionality in Fig. 6.1 is about 109/100 = 1.09, but depends on the size of the specified ripples in the passband and stopband. To get a useful result we need to find out how this slope depends on the ripples.

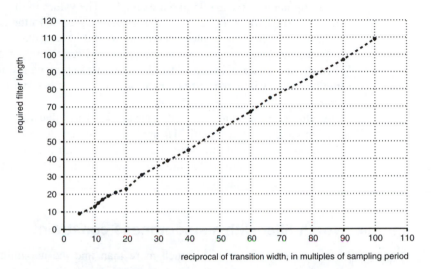

Fig. 6.1 Some empirical data: required feedforward filter length as a function of the reciprocal of transition width. The lower edge of the stopband in the example of Section 5 is varied, and all the other specifications are kept the same.

Well, the same sort of experiment can be carried out by varying the passband ripple and keeping everything else fixed, then varying the stopband ripple and keeping everything else fixed. It turns out that when the transition width is kept fixed, the filter length is roughly proportional to the logarithm of the ripple specification in either the passband or stopband.

James Kaiser found a particularly elegant way to estimate the required filter length (see the Notes for the reference). To write it very succinctly, we need to talk about a good way to express transition width and ripple. Let ΔF be the transition width in fractions of the sampling rate — that's simple. Expressing the ripple is a bit more complicated.

Suppose we denote the ripples in the passband and stopband by δ_1 and δ_2, respectively. (In our example, $\delta_1 = \delta_2 = 0.05$.) Then the geometric mean of the two ripples is $(\delta_1 \delta_2)^{1/2}$. (In our case the geometric mean is just 0.05, of course.) We said the required filter length is roughly proportional to the logarithm of the ripple, so it's convenient to express the ripple in dB. Ripple is normally less than 1, corresponding to negative dBs, so it's even more convenient to use the negative of the dB measure. So, finally, we'll define the ripple in dB to be

$$R = -20 \log_{10} (\delta_1 \delta_2)^{1/2} \tag{6.1}$$

In our example, $R = 26.02$ dB. (Check this on your calculator.)

With these definitions of ΔF and R, Kaiser's formula is

$$n \approx \frac{R - 13}{(14.36) \Delta F} + 1 \tag{6.2}$$

Very compact. But let me warn you that this is meant to serve only as a rough guide for a very wide range of design parameters. It's more accurate for ripple specifications that are tighter than the ± 0.05 in our example. The values in that example, $R = 26.02$ and $\Delta F = 0.04$, yield $n \approx 23.7$, which is not very close to the actual minimum filter length $n = 31$. But this ripple specification of ± 0.05 is actually quite large as ripples go. In fact, I chose it that large just so you could see the ripple clearly in the graphs. When the specified ripple is ± 0.001 ($R = 60$ dB) in both bands, for example, with the same transition width, the formula gives an estimate of $n \approx 81.4$, and the true value is $n = 83$.

At this point in your study of filtering, knowing that Eq. 6.2 exists is probably more important than any particular application to design problems. After all, you can always experiment with programs like METEOR. But it adds to your general education to know that required feedforward filter length is close to being proportional to ripple in dB, and, at the same time, inversely proportional to transition width.

7 Example: Notch filter with a smoothness constraint

Linear programming can do much more than find the maximin approximation to a given frequency response. For one thing, we don't have to push the frequency response away from every constraint. To allow the frequency response to hug a constraint, all we need to do is omit the δ in Eqs. 4.3 and 4.4.

We can also put constraints on the derivatives of the frequency response. For example, it's often nice to be able to stipulate that the response be convex up or down in some band. To see this, just notice that if we differentiate the cosine terms like $c_i \cos(i\omega)$ in the frequency response, we get terms of the form $-ic_i \sin(i\omega)$. If we differentiate again, we get terms of the form $-i^2 c_i \cos(i\omega)$, and so on. For each fixed value of ω on the frequency grid, a specification on a derivative will simply yield another constraint on the unknown coefficients. What's crucial is that these constraints are *linear* in the coefficients, just as the original constraints are.

There are even more general things we can do with linear programming, such as putting direct constraints on the coefficients, which after all represent the impulse response. It's important to realize just how general the approach really is, so let's look at two more examples.

Figure 7.1 shows the result when we ask for an odd-length filter with a lower and upper passband and a linear notch between them. By this I mean a portion of the frequency response that descends from unity in a straight line to 0, and then returns, also linearly. We gild the lily by requiring that the response in the first half of the lower passband be convex down, which will make it quite smooth at low frequencies. The specification to METEOR actually has the following nine constraint specifications:

- upper and lower bounds in lower passband;
- upper and lower bounds in upper passband;
- upper and lower bounds in descending part of notch;
- upper and lower bounds in ascending part of notch;
- convexity constraint in first half of lower passband.

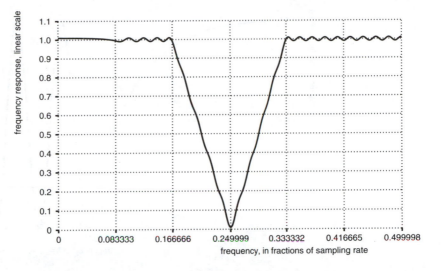

Fig. 7.1 Illustrating the flexibility of METEOR. This filter is specified as having passbands for the first and last third of the baseband, and a linear notch in the second third. Also, the frequency response in the first half of the lower passband is constrained to be convex down. With tolerances of ±0.01, we need a length-111 feedforward filter. Note the linear vertical scale to show the linearity of the notch.

The order in which the constraints are given is immaterial. The linear programming algorithm doesn't mind if they're scrambled, as long as they're all there.

Figure 7.2 shows a close-up of the lower passband, showing the effect of the convexity constraint. This is an easy way to ensure flatness in particular bands. Don't be alarmed by the wild variations in frequency response; the scale is blown up and represents a deviation of only ±1 percent. By the way, a convexity constraint like this one need not cost much in terms of filter length. The minimum length meeting these nine constraint specifications is 111 coefficients, but removing the convexity constraint reduces the required length only to 109.

8 Example: Window design

The second example shows how METEOR can be used to design a window for spectrum measurement. Recall from Chapter 10 that the measured spectrum of a signal is smoothed by convolution with the transform of the window. Designing a particular window frequency content for this smoothing is exactly the same problem as designing a feedforward filter; the coefficients are the window sample values. What we want is a frequency content that is unity at zero frequency and that descends to small values as quickly as possible. In the terminology of window design, we want a narrow central lobe and low side lobes.

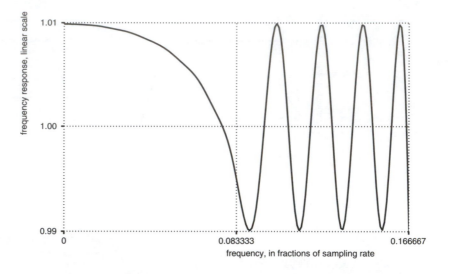

Fig. 7.2 Close-up of the first passband in the previous figure, with greatly
expanded vertical scale.

How do we choose specifications to achieve this result? The first thing to notice is
that the length of the window is fixed by the length of the FFT we're using. That
means that the linear programming algorithm will be used with a fixed number of
coefficients, and no search is required to find the minimum length that satisfies our
constraints. (I know I said we would restrict our attention to odd-length filters, but
we'll choose a length of 64 in this example because we're designing an FFT window,
and FFTs most often use lengths that are powers of 2.)

The next thing to realize is that, with the length fixed, we are left with two design
parameters: the depth of the side lobes and the width of the central lobe. We can fix
either one and optimize the other. Both strategies are in the repertoire of METEOR,
and for this example we'll fix the depth of the side lobes (at 60 dB rejection, or
±0.001), and minimize the width of the central lobe. METEOR does this by pushing
the left edge of the stopband as far to the left as possible while keeping the stopband
response down at the minimum specified rejection.

The final obstacle to overcome is fixing the response at zero frequency to unity.
But we've already mentioned this option above — all we need to do is omit the
"squeezing" parameter δ, and let the response hug the following constraints at zero
frequency:

$$\hat{H}(0) \leq 1 \quad \text{and} \quad \hat{H}(0) \geq 1 \tag{8.1}$$

which are, of course, equivalent to $\hat{H}(0) = 1$. This particular "band" of frequencies
consists of only one point. By omitting the parameter being optimized, we are thus
able to pin down the response at a particular frequency to a specific value.

So the complete problem statement is to find a length-64 filter, with 60 dB rejection in the stopband and the narrowest possible central lobe. There are actually the following four band specifications to METEOR:

- the two constraints in Eq. 8.1, setting $\hat{H}(0) = 1$;
- the upper and lower bounds in the stopband (±0.001).

I started with a stopband edge for which 60 dB rejection is achievable. I found such an edge by trial and error, using METEOR to maximize rejection for a few fixed band edges. It turns out that 0.04 times the sampling frequency does the trick, yielding 62.9 dB rejection. I therefore have an extra 2.9 dB rejection to play with, which makes it possible to slide the band edge a little to the left. METEOR tells me just how far, and gives me a final design with almost precisely 60 dB rejection and a final band edge that is 0.03831 times the sampling frequency.

The resulting frequency response — or what in this case should be called window frequency content — is shown in Fig. 8.1. The corresponding frequency content for the Hamming window, Fig. 5.2 of Chapter 10, doesn't have such nice uniform rejection in the stopband. Howard Helms (see the Notes) puts it perfectly: this window has "the best possible resolution for a given maximum leakage."

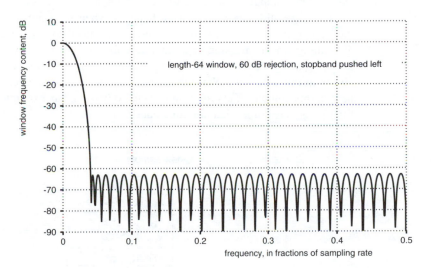

Fig. 8.1 Frequency content of a length-64 window, designed with 60 dB rejection in the stopband, and the lower end of the stopband pushed left as far as possible. The resulting stopband starts at 0.03831 times the sampling frequency.

Brawn can sometimes substitute for brains (but not often). It turns out that the problem of designing windows satisfying this particular criterion of optimality has been solved by sheer intellectual power. It was done back in 1946 by C. L. Dolph, building on ideas of Chebyshev — and they're called *Dolph-Chebyshev* windows. The result in Fig. 8.1 is exactly what you get if you use Dolph's formula. And algorithms for solving linear programming problems hadn't even been invented in 1946. But having a design program as general as METEOR puts us ahead of the game, and today we can design windows with different rejection in different parts of the stopband, with arbitrary shapes to the central lobe, and practically any other weird requirement we can dream up.

9 A programming consideration

We ought to take a look at how feedforward filters are actually implemented on a computer. There's a smart way to do it and a not-so-smart way, and explaining both gives me a chance to illustrate a general and useful idea from computer science.

Suppose for the purposes of discussion we agree to implement a length-8 filter. To compute the output at time t, we need to have available the input samples $x_t, x_{t-1}, \ldots$ down to x_{t-7}. It's very hard to think of doing anything else but storing them in an array, a sequence of consecutive storage locations. The required weighted summation,

$$y_t = a_0 + a_1 x_{t-1} + a_2 x_{t-2} + \cdots + a_{n-1} x_{t-(n-1)} \tag{9.1}$$

can then be computed conveniently in a simple loop.

The interesting question is what happens when we want to move ahead and find the output at time $t+1$. Figure 9.1 shows the most obvious thing to do. Just move everything down one slot, making room for the newly arrived input sample, and throwing away the oldest stored sample, which will never be needed again. The problem with this solution is that we move every piece of data every time we compute a new output sample. This might not seem so bad for our example of length eight, but filter lengths ten times that are more common.

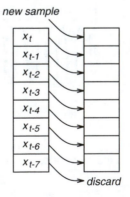

Fig. 9.1 Updating our array of present and past input samples for feedforward filtering by moving everything one place. Not so smart.

Moving all the data every time a new point arrives is obviously a bad idea, and the way around the problem is to change what we regard as the beginning of the array instead of moving everything. In computer science terms, we use a pointer to tell us the location of the present sample, x_t. The past samples are below that point in the array, and use up all of the remaining space, wrapping around when we get to the end of the array. (Recall that the time and frequency domains of the DFT are circular arrays. This idea should be very familiar by now; if you're at all unsure of it, reread the beginning of Chapter 8.)

For example, suppose we are in the situation where we regard the beginning of the array as being in position 2. Then the present sample, x_t, is in position 2; x_{t-1} is in position 1; and x_{t-2} is in position 0. To get x_{t-3}, we need to wrap around to position 7, and work our way down from there, finally getting to x_{t-7} in position 3.

Figure 9.2 shows the same array as Fig. 9.1 but in circular form. On the left we see what happens when a new sample arrives, after we've computed the tth output value. The place marked now holds the most recent input sample. The value x_{t-7} will never be needed again, so its place is the natural place — in fact, the only place — to put the new sample value. If we next just add one to now, so that it points to the newly arrived sample, we're all set to compute the next output value, as shown on the right. What was x_t is now x_{t-1}, and so on. We've had to change only one element in the array.

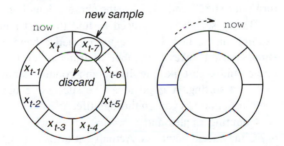

Fig. 9.2 Updating our array of input samples by changing where we start. The circular array makes it unnecessary to move the data. The array positions are numbered so that they increase in the clockwise direction.

The steps I've just described are very easy to translate into code. Suppose the filter length is L and the present and past input samples are stored in array[i], where i ranges from 0 to L-1. Then the following piece of code does the trick:

```
now = now + 1;
if ( now > L-1 ) now = now - L;
array[now] = new_sample;
```

At a given time t, the filtering operation indexes through every element of the array, wrapping around as it did above for the example of a length-8 filter.

So the circular array gives us an efficient way to program the filters that linear programming produces from a design program like METEOR. You now know how to go from specifications in the frequency domain to a set of coefficients for a feedforward filter to a computer program that actually implements the filter. This is what I meant when I claimed at the beginning of the chapter that the feedforward design problem is essentially solved, at least as far as our everyday needs are concerned. In the next chapter we'll take a look at the analogous process for feedback filters.

Notes

The write-up of METEOR has already been referenced at the end of Chapter 4. There's no magic to the way it finds the shortest-length filter that satisfies given specifications. It simply uses binary search, and repeatedly looks for solutions to linear programs. For example, it may start with a given range of filter lengths from 1 to 127. METEOR first tries the midpoint, or, more precisely, an odd length closest to the midpoint, say 63. If there is a filter of length 63 that satisfies the constraints, it restricts the range between 1 and 63; if not, it restricts the range between 65 and 127. It then repeats the process until it finds two successive odd lengths, with the filter feasible for the higher of the two and infeasible for the lower. For example, if those two lengths turn out to be 29 and 31, 31 is the shortest length that satisfies the constraints.

Binary search is also used to find the best band edge, as in the window design example in Section 8, where we pushed a band edge to the left as far as possible. Binary search takes a number of trials proportional to the logarithm of the initial range, and is just like the divide-and-conquer strategy described in Chapter 8 for the FFT and sorting. A good idea like successive binary subdivision goes a long way. Keep it in mind as you go through life.

Counting ripples and worrying about when there are $m+1$ and when there are $m+2$ may seem pointless. Actually, it turns out that knowing the number of ripples in an optimal design response is the key to finding the very efficient design algorithm of Parks and McClellan (also mentioned in the Notes to Chapter 4).

The closed-form windows called Dolph-Chebyshev were described in

> H. D. Helms, "Nonrecursive Digital Filters: Design Methods for Achieving Specifications on Frequency Response," *IEEE Trans. Audio & Electroacoustics*, vol. AU-16, no. 3, pp. 336–342, Sept. 1968.

and Dolph's paper is

> C. L. Dolph, "A Current Distribution for Broadside Arrays which Optimizes the Relationship Between Beam Width and Side-Lobe Level," *Proc. IRE*, vol. 35, pp. 335–348, June 1946.

The simplex algorithm for solving linear programming, which is used by METEOR, was actually invented by George Dantzig in 1947, the year after Dolph's paper appeared. Computers that were fast enough to solve linear programming problems like the ones in this chapter were not commonly available for another 30 years or so.

Kaiser gave his formula for estimating the required length of feedforward filters in

J. F. Kaiser, "Nonrecursive Digital Filter Design using the I_0-*SINH* Window Function," *Proc. 1974 IEEE Int. Symp. on Circuits and Systems,* pp. 20–23, April 1974. (Reprinted in *Selected Papers in Digital Signal Processing II*, Digital Signal Processing Committee of the IEEE ASSP Society (eds.), IEEE Press, New York, N.Y., 1975.

The title of the paper refers to a method for designing feedforward filters using a windowed version of the Fourier series for the frequency response. After all, as Eq. 2.9 shows, the frequency response of a feedforward filter is a partial Fourier series. Since a finite number of terms is used, the actual response ripples a lot, so the coefficients should be windowed. This approach doesn't give answers that are optimal — in the sense that iterative methods like METEOR do — but the algorithm is easy to understand, and fast to code and run. The "I_0-*SINH*" refers to a particularly useful class of windows, now called *Kaiser windows*.

Notice that what I call feedforward filters were then called "nonrecursive." Today, they're usually called "FIR" filters, as mentioned in the Notes to Chapters 4 and 5.

Good sources for more information about digital filter design in general are the handbook

S. K. Mitra and J. F. Kaiser (eds.), *Handbook for Digital Signal Processing*, John Wiley, New York, N.Y., 1993,

and

T. W. Parks and C. S. Burrus, *Digital Filter Design*, John Wiley & Sons, New York, N.Y., 1987,

as well as the following two standard texts:

A. V. Oppenheim and R. W. Schafer, *Digital Signal Processing*, Prentice-Hall, Englewood Cliffs, N.J., 1975.

L. R. Rabiner and B. Gold, *Theory and Application of Digital Signal Processing,* Prentice-Hall, Englewood Cliffs, N.J., 1975.

Problems

1. For some problems it's possible to prove that iterative techniques are absolutely necessary. Finding the roots of polynomials is one such example. For what degree polynomials is there no closed-form solution? If you don't happen to know the answer, look it up. Proving this was one of the great achievements of humankind.

2. Work out the form of the frequency response analogous to Eq. 2.9 when the feedforward filter length is even.

3. When is the frequency response of a feedforward filter a sine instead of a cosine series?

4. What does the symmetry in the frequency response for the example in Fig. 5.1 imply about the filter coefficients?

5. Formulate a heuristic argument that shows that the optimal feedforward filter design is equiripple, using geometry like that shown in Fig. 4.1.

6. For the purposes of the length-estimation formula in Eq. 6.2, the passband ripple in dB is $20\log_{10}\delta$, where δ is the deviation from nominal specification. Sometimes the passband ripple is expressed in dB by $20\log_{10}[(1+\delta)/(1-\delta)]$. Why?

7. Count the number of points at which the frequency response ties for distance from the constraints in the half-band example in Fig. 5.1. These are all counted as ripples. Don't forget to count the edges of the bands — the frequency response does just hit the maximum deviation at both edges of both the passband and stopband. I stated at the end of Section 5 that the number of ripples is always one or two more than the number of free coefficients, m. Does this example satisfy my claim? Which is it in this case, $m+1$ or $m+2$?

8. Why are the bottom edges of the dB plots like Fig. 5.2 often ragged? What would you do to make them uniform?

9. Suppose we've decided on a set of specifications for a feedforward filter design. For each odd filter length n, there either is or isn't a set of coefficients that result in a feasible design. We can think of this as defining, for the particular specifications, a function $F(n)$ that takes on the value "feasible" or "infeasible" for each odd integer value of the variable n. What property of the function $F(n)$ ensures that binary search always successfully finds the smallest odd value of n for which there is a feasible filter?

10. Write code that actually does feedforward filtering using the circular array discussed in Section 9. How should it be initialized?

11. The coefficients of the feedforward filters discussed in this chapter are symmetrical about their center. Explain how to take advantage of this to almost halve the number of multiplications needed to do the filtering. Rewrite the code in the previous problem to achieve this.

CHAPTER 13

Designing Feedback Filters

1 Why the general problem is difficult

Simplex for linear programming is an iterative optimization algorithm, and as we saw in the previous chapter, it works very well for designing quite arbitrary feedforward filters. Why doesn't the same approach work for feedback filters? The answer is simple: the corresponding design problems for feedback filters are not linear. In fact they're very nonlinear, which means trouble for iterative numerical methods. That's why we'll be content with closed-form designs for the commonly used feedback filters — lowpass, highpass, and bandpass. Fortunately, these cover most important applications.

To get a little more insight into what the difficulty is, consider first the real-valued frequency response of a centered linear-phase feedforward filter like the ones we designed in Chapter 12:

$$\hat{H}(\omega) = c_0 + c_1\cos\omega + c_2\cos(2\omega) + \cdots + c_m\cos(m\omega) \qquad (1.1)$$

(Remember that we're using symmetric-coefficient filters, so that the frequency response without the linear-phase factor is always a real number.) METEOR determines the best choices for the coefficients by adjusting them so that this response best approximates a desired response. It's important that the response changes in a reasonable way when the coefficients are varied. If changing a coefficient has a weird effect on $\hat{H}(\omega)$, it's going to be hard to find optimal filters.

The important point is that if any one coefficient, say c_i, is varied, and all the others are held fixed, the change in $\hat{H}(\omega)$ is proportional to c_i. In other words, the frequency response of a feedforward filter is a *linear* function of each coefficient.

By way of contrast, the transfer function of a feedback filter is of the form

$$\mathcal{H}(z) = \frac{a_0}{1 + b_1 z^{-1} + b_2 z^{-2} + \cdots + b_k z^{-k}} \tag{1.2}$$

and the magnitude of the frequency response is the magnitude of this when $z = e^{j\omega}$. We see immediately that we're in for trouble when we start moving around the denominator coefficients to achieve an optimal design. The frequency response depends on the coefficients b_i in a much more complicated way than in the feedforward case. For example, there are sets of coefficients that cause the frequency response to be infinite at some points, which is never the case for feedforward filters. Even worse, if the coefficients are such that there's a pole outside the unit circle, the filter will be unstable and useless.

So optimization algorithms for feedforward filters see gently rolling plains. But the landscape for feedback filters is terrifying, riddled with spires and crannies. It's easy to believe that the general feedback filter design problem, with twenty or thirty coefficients to optimize simultaneously, can be very nasty.

2 The Butterworth frequency response

Having given up on the idea of a general design program (but see the Notes), I'm now going to derive a very useful class of closed-form feedback filters from the ground up. These will turn out to be the *Butterworth* filters, the simplest kind of closed-form filters. Other, more complicated filters are designed using exactly the same ideas.

Let's start with the problem of designing a lowpass digital filter. What we're after is a transfer function that has both zeros and poles, a ratio of polynomials in the frequency variable z. At zero frequency, where $z = 1$, we want the transfer function to be, say, one; and at the Nyquist frequency, where $z = -1$, we want the transfer function to be zero. Furthermore, we want the transition from one (in the passband) to zero (in the stopband) to be as abrupt as possible, for a given number of terms in the numerator and denominator.

If we play around with the very simplest ratios of polynomials, we find that the function

$$\mathcal{H}(z) = \frac{1}{1 + \dfrac{z-1}{z+1}} \tag{2.1}$$

seems to satisfy the requirements. Just check: When $z = 1$, the second term in the denominator is zero, so $\mathcal{H}(1) = 1$. When $z = -1$, that term is infinite and is in the denominator, so $\mathcal{H}(-1) = 0$. This is not a totally crazy lowpass filter. In fact, it actually turns out to be a feedforward filter, and we've seen it before (see Problem 2). There's a problem though, because the $(z-1)/(z+1)$ term in the denominator takes on complex values as z travels around the frequency circle. This makes it difficult to control the magnitude of the frequency response, which is our ultimate aim. But we've only just begun, and we can do a lot better.

As I've just mentioned, we're really interested in controlling the *magnitude* of $\mathcal{H}(z)$ when z is on the unit circle, $|H(\omega)|$. To do this, we'll use a trick that is useful in other contexts as well. Suppose you start with any real function of the complex frequency variable z, say $F(z)$. By a "real function" I mean that there are no js explicitly in F's definition, so that when z is real, $F(z)$ is also real. Consider the product $F(z)F(z^{-1})$. This new function has some very interesting properties. First, it's symmetric in z and z^{-1}, meaning that replacing z by z^{-1} has no effect at all. Thus, if it has a pole or zero at some point $z = p_0$, it must also have a corresponding pole or zero at the reciprocal point p_0^{-1}. This implies that every pole or zero inside the unit circle has a corresponding image outside the unit circle.

Next, consider the values of $F(z)F(z^{-1})$ when z is on the frequency circle, $z = e^{j\omega}$. The values of $F(e^{-j\omega})$ are the complex conjugates of the values of $F(e^{j\omega})$, simply because j is everywhere replaced by $-j$. (Remember that F is a real function.) Therefore, on the frequency circle, $F(z)F(z^{-1})$ is equal to the squared magnitude of $F(z)$, and is real and non-negative.

This property is just what we need to control the behavior of the frequency response $\mathcal{H}(z)$ in Eq. 2.1. Let our function $F(z)$ be the term $(z-1)/(z+1)$ in the denominator of that equation, and rewrite $\mathcal{H}(z)$ as

$$\mathcal{H}(z) \;=\; \frac{1}{1 \,+\, F(z)} \tag{2.2}$$

If we now replace $F(z)$ by $F(z)F(z^{-1})$, which is real and non-negative for all frequencies, the new denominator will be real and will go smoothly from one to infinity as the frequency increases from zero to the Nyquist frequency. I'm therefore proposing that we use the function

$$\frac{1}{1 \,+\, F(z)\,F(z^{-1})} \tag{2.3}$$

An even better idea is to make the transition from one to infinity faster by raising the real product $F(z)F(z^{-1})$ to a power, resulting in the function

$$\frac{1}{1 \,+\, [F(z)\,F(z^{-1})]^N} \tag{2.4}$$

This does have precisely the effect of making the transition of the denominator from one to infinity more abrupt, because raising something to the Nth power makes it smaller when it's smaller than one, and larger when it's larger than one. We'll see below that our final result will be a feedback filter with N poles. The larger N is, the sharper the cutoff.

Except for one detail, this, finally, is the transfer function of a Butterworth filter. The problem that remains is that the transfer function has poles outside the unit circle, and is therefore unstable. As it stands, the corresponding filter is unusable. To see why, observe that Eq. 2.4 is symmetric in z and z^{-1}. By the same reasoning as above, each pole inside the unit circle has a corresponding pole outside. The solution to this problem is to use only the poles inside the unit circle. We'll also use only the zeros inside the unit circle, for reasons you don't have to worry about right now (see Problem 3).

To get the final, stable transfer function, collect all the factors associated with poles and zeros of the function in Eq. 2.4 that are inside the unit circle, multiply them together, and call the result $B(z)$. By symmetry, the rest of the transfer function must be $B(z^{-1})$, and we can rewrite Eq. 2.4 as

$$B(z)\, B(z^{-1}) = \frac{1}{1 + [F(z)\, F(z^{-1})]^N} \qquad (2.5)$$

The function $B(z)$ is then the desired transfer function, and is stable because all of its poles are inside the unit circle.

We can now, at last, enjoy the fruits of our work, and take a look at the frequency response of these famous and useful filters. Notice that evaluating Eq. 2.5 on the frequency circle actually results in the squared magnitude of the frequency response of $B(z)$, as we discussed above when we formed $F(z)\, F(z^{-1})$. The second term in the denominator can be evaluated easily enough if we remember that $F(z) = (z-1)/(z+1)$ and let $z = e^{j\omega}$:

$$F(e^{j\omega}) = \frac{e^{j\omega} - 1}{e^{j\omega} + 1} \qquad (2.6)$$

This should be a familiar situation. Multiply the top and bottom by $e^{-j\omega/2}$, getting

$$F(e^{j\omega}) = j \tan(\omega/2) \qquad (2.7)$$

Substituting this in Eq. 2.5 then yields the squared magnitude

$$|B(\omega)|^2 = \frac{1}{1 + \tan^{2N}(\omega/2)} \qquad (2.8)$$

This has just the properties we've been aiming for. When $\omega = 0$, $\tan(\omega/2)$ is zero, and the frequency response is one. When $\omega = \pi$ radians per sample (the Nyquist frequency), $\tan(\omega/2)$ is infinity, and the frequency response is zero. Figure 2.1 shows the entire frequency response for $N = 4$, 8, and 16, and as you can see, these are quite respectable half-band filters. In Problem 5 you'll see that it's easy to determine how high a value of N is required to meet certain specifications.

The frequency at which the cutoff occurs is exactly the frequency at which the tangent in the denominator of Eq. 2.8 is one; for lower frequencies that term becomes smaller, and for larger frequencies it becomes larger. That frequency is, of course, $\omega = \pi/2$ radians per sample, half the Nyquist frequency, so in fact these Butterworth filters are all approximations to half-band filters. Later in this chapter we'll see how to shift the cutoff frequency.

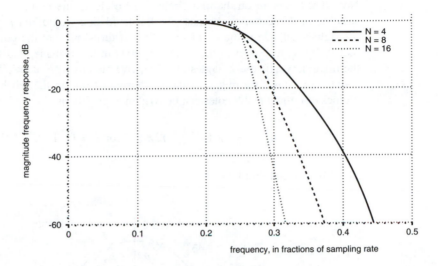

Fig. 2.1 The frequency response of the Butterworth filters of order 4, 8, and 16. Note the dB scale.

3 The Butterworth poles and zeros

We now have the Butterworth filter frequency response. In this section we're going to determine where its poles and zeros are; then we'll know its transfer function completely and be able to implement it.

Go back to Eq. 2.5:

$$\mathcal{B}(z)\,\mathcal{B}(z^{-1}) = \frac{1}{1 + [F(z)\,F(z^{-1})]^N} \tag{3.1}$$

and recall that $F(z) = (z-1)/(z+1)$. It's easy to see that $F(z^{-1}) = -F(z)$, so we can rewrite this as

$$\mathcal{B}(z)\,\mathcal{B}(z^{-1}) = \frac{1}{1 + (-F^2)^N} \tag{3.2}$$

We can now find the poles in terms of the function F; the rest will be easy. The poles occur where the denominator is zero, and so are determined by the equation

$$1 + (-1)^N F^{2N} = 0 \tag{3.3}$$

Multiplying by $(-1)^N$ and rearranging puts this in a more familiar form:

$$F^{2N} = (-1)^{N+1} \tag{3.4}$$

Thus, the poles occur in the complex F-plane at the $(2N)$th roots of $(-1)^{N+1}$, which are equally spaced around the unit circle, at angular increments of $2\pi/(2N)$ radians. Note that the point on the unit circle at 12 o'clock, the point $F = j$, can never be a solution of this equation, because the left-hand side would be $(j)^{2N} = (-1)^{N}$, while the right-hand side is the negative of that. It turns out that the poles are placed symmetrically with respect to the imaginary axis in the F-plane, and the poles closest to the imaginary axis make angles of $\pm 2\pi/(4N)$ radians from it. We'll number the poles counterclockwise, starting with the northernmost one in the left-hand F-plane (at 11 o'clock), so that the $2N$ poles can be written explicitly as

$$\theta_i = \frac{\pi}{2} + \left(\frac{2i+1}{4N}\right)2\pi , \quad \text{for } i = 0, 1, \ldots, 2N-1 \tag{3.5}$$

(See Fig. 3.1 and Problem 6.)

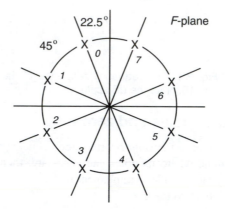

Fig. 3.1 The eight poles in the F-plane when developing the four-pole half-band Butterworth filter. The northernmost poles lie at angles $\pm 2\pi/(4N)$ radians = 22.5° from the imaginary axis. The poles are numbered in accordance with Eq. 3.5.

Next, we're going to find the location of the poles in the z-plane. This is no problem, because we know F in terms of z:

$$F = \frac{z - 1}{z + 1} \tag{3.6}$$

and we can solve this for z in terms of F:

$$z = \frac{1 + F}{1 - F} \tag{3.7}$$

We've just seen that the poles of F are on the unit circle in the F-plane, so let $F = e^{j\theta_i}$. Then we have almost the same situation as we did in Eq. 2.6; the result this time is that

$$z = \frac{j}{\tan(\theta_i/2)} = jp_i \tag{3.8}$$

So we see that the poles in the z-plane lie on the imaginary axis. This makes perfect sense because we're dealing with a half-band filter — the cutoff frequency is exactly halfway between zero and the Nyquist frequency, so the poles lie on the axis of symmetry separating the points $z = 0$ and $z = \pi$ radians per sample. From now on we'll use the shorthand notation $p_i = 1/\tan(\theta_i/2)$ for the pole locations on the imaginary z-axis.

Take a moment to observe the very convenient properties of the transformation represented by the function $F(z)$. In the previous section we evaluated the frequency response of the Butterworth filters, and saw in Eq. 2.7 that when $z = e^{j\omega}$, $F(z)$ was purely imaginary. Now we see that the opposite is also true: when F is on the unit circle in the F-plane, z is purely imaginary. That is, the unit circle in each plane is the image of the imaginary axis in the other.

As explained in the previous section, we're going to use only those poles of $\mathcal{B}(z)\,\mathcal{B}(z^{-1})$ that lie inside the unit circle in the z-plane, the ''stable'' poles. We now know exactly where those poles occur, so it's a simple matter to select the stable ones. Equation 3.8 tells us that for each pole in the F-plane at the point θ_i on the unit circle, there is a pole in the z-plane at the point $p_i = 1/\tan(\theta_i/2)$ on the imaginary axis. Therefore the stable poles in the z-plane are precisely those for which the tangent is greater than one in magnitude, and these correspond to the angles $\pi/2 < \theta_i < 3\pi/2$. Thus, the poles in the left-half F-plane are the ones that show up inside the unit circle in the z-plane, the first N of them by the numbering scheme in Eq. 3.5.

We now know everything we need to know to write the complete, stable transfer function of the half-band Butterworth filter with N poles in terms of its poles and zeros. Replacing $F(z)$ explicitly by $(z-1)/(z+1)$, Eq. 3.1 becomes

$$\mathcal{B}(z)\,\mathcal{B}(z^{-1}) = \frac{1}{1 + \left[-\left(\frac{z-1}{z+1}\right)^2 \right]^N} \tag{3.9}$$

Multiplying the top and bottom by $(z+1)^{2N}$ shows that there are $2N$ zeros at the point $z = -1$, half of which will show up in $\mathcal{B}(z)$.

The stable poles are in the left-half F-plane, and are therefore indexed in Eq. 3.5 by $i = 0, 1, \ldots, N-1$. It's now convenient to assume that N is even (I'll leave the case when N is odd for Problem 7). In the N-even case there is no single, leftover pole at $z = -1$, and the N poles occur in complex pairs, so that we need to worry only about the first $N/2$. Each pole at jp_i has a complex conjugate partner, and those two poles combine to form a factor $z^2 + p_i^2$. The transfer function is therefore

$$\mathcal{B}(z) = \frac{A(z+1)^N}{(z^2 + p_0^2) \cdots (z^2 + p_{N/2-1}^2)} \tag{3.10}$$

The constant factor A is there just to control the overall level of the output, and is usually called the *gain constant*. It's arbitrary, but it's often convenient to set the gain to unity at zero frequency, in which case we can compute A from the condition $\mathcal{B}(1) = 1$ (see Problem 4).

We have the transfer function of the half-band Butterworth filter with N poles, and furthermore, we have factored it in terms of known poles and zeros. We choose the number of poles N large enough to get the desired passband flatness, sharpness of cut-off, and stopband rejection. This is easy because we have a simple formula for the magnitude of the frequency response (see Eq. 2.8 and Problem 5).

But what if we want the cutoff frequency at some other point? What if we want a highpass filter? We'll see in the next two sections that these questions have simple answers, now that we've derived the basic prototype transfer function.

4 More general specifications

Suppose now that we want the cutoff frequency of a Butterworth filter to be ω_c radi-ans per sample instead of $\pi/2$ radians per sample. We can make this happen by going back to the squared-magnitude frequency response in Eq. 2.8:

$$|B(\omega)|^2 = \frac{1}{1 + \tan^{2N}(\omega/2)} \tag{4.1}$$

At the cutoff frequency $\omega = \pi/2$ radians per sample, the second term in the denomi-nator has the value one, and the squared-magnitude frequency response always has the value $1/2$ at that point, no matter what the value of N is. All we need to do to shift the cutoff frequency to ω_c is to make that term have the value one at $\omega = \omega_c$ instead of at $\omega = \pi/2$. That's easy: just replace $\tan(\omega/2)$ by $\tan(\omega/2)/\tan(\omega_c/2)$:

$$|B(\omega)|^2 = \frac{1}{1 + \left[\tan(\omega/2)/\tan(\omega_c/2)\right]^{2N}} \tag{4.2}$$

To make the derivation go through in exactly the same way as before, simply change the definition of $F(z)$ in Eq. 2.5 to

$$F(z) = \frac{1}{\tan(\omega_c/2)} \cdot \frac{z-1}{z+1} \tag{4.3}$$

Equation 3.2 remains exactly the same:

$$\mathcal{B}(z)\,\mathcal{B}(z^{-1}) = \frac{1}{1 + (-1)^N F^{2N}} \tag{4.4}$$

and the poles θ_i in the F-plane are unchanged.

We can now find the poles in the z-plane as before, solving Eq. 4.3 for z in terms of F:

$$z = \frac{1 + F \cdot \tan(\omega_c/2)}{1 - F \cdot \tan(\omega_c/2)} \tag{4.5}$$

Just to check, when $\omega_c = \pi/2$ everything we've just done reduces to the half-band case, as it should. We know the values of F at the poles — they're on the unit circle at the angles θ_i in Eq. 3.5, so the poles in the z-plane now occur at the $2N$ points

$$z_i = \frac{1 + e^{j\theta_i} \cdot \tan(\omega_c/2)}{1 - e^{j\theta_i} \cdot \tan(\omega_c/2)}, \quad \text{for } i = 0, 1, \ldots, 2N-1 \tag{4.6}$$

Figure 4.1 shows the poles in the z-plane when the cutoff frequency is $2\pi/10$ radians per sample, one-tenth the sampling rate. The corresponding frequency response is shown in Fig. 4.2. The poles have now moved off the imaginary axis, where they were for the half-band case, and have become squeezed toward the zero-frequency point, $z = 1$. As we might expect, the lower the cutoff frequency, the more the poles will move toward the zero-frequency point.

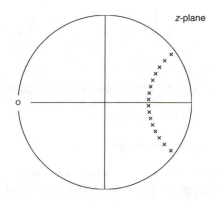

Fig. 4.1 The poles and zeros in the z-plane for a 16-pole lowpass Butterworth filter with cutoff frequency at one-tenth the sampling rate. The circle is, of course, the unit circle. The zero shown at $z = -1$ actually represents 16 zeros at the same spot, as indicated in Eq. 4.7.

The zeros, on the other hand, stay in the same place, $z = -1$, for the same reason as before: The factor $(z+1)^{2N}$ shows up in the numerator of $\mathcal{B}(z)\,\mathcal{B}(z^{-1})$ when it is cleared of fractions, as in Eqs. 3.9 and 3.10. Half of those zeros belong to $\mathcal{B}(z)$, and the other half to $\mathcal{B}(z^{-1})$.

The poles in the z-plane are now no longer necessarily on the imaginary axis, so we write the transfer function in the general form

$$\mathcal{B}(z) = \frac{A(z+1)^N}{(z - z_0) \cdots (z - z_{N-1})} \tag{4.7}$$

As before, the gain factor A is usually chosen to make the gain unity at zero frequency.

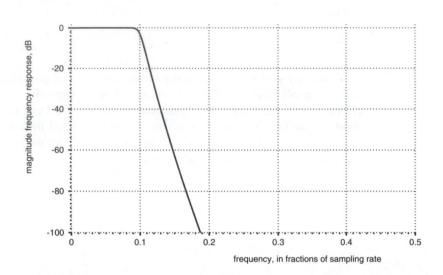

Fig. 4.2 The frequency response of the 16-pole Butterworth filter whose poles and zeros are shown in Fig. 4.1. The cutoff frequency is now one-tenth the sampling frequency.

5 A lowpass/highpass flip

We just derived the Butterworth lowpass filter in two steps. First, we found a very special prototype — the half-band filter in Eq. 3.10. Then we transformed it to get an arbitrary cutoff frequency. The transformation step was simple: to move the cutoff frequency from half the Nyquist to ω_c, we just divided the function F by the constant factor $\tan(\omega_c/2)$, as in Eq. 4.3.

This two-step process is an example of the strategy commonly used for designing closed-form feedback filters. We start with special prototypes and transform them to more generally useful forms. I'll illustrate the idea again with a very useful trick.

Return to the question of designing a *highpass* Butterworth filter. Think for a moment about what we need to do to convert a lowpass filter into a highpass filter. Somehow, we ought to interchange the high and low frequencies. We'd like to make zero frequency correspond to the highest possible frequency, the Nyquist, and vice versa. We'd also like the frequencies between the two to be in reverse order, so that lower frequencies correspond to higher.

What does this mean in terms of the frequency variable z? Well, the zero and Nyquist frequencies correspond to the points $z = 1$ and $z = -1$, respectively. The simplest thing in the world would be to replace z by $-z$ in the transfer function — which would at least ensure that the lowest and highest frequencies are interchanged. That sounds too simple to work, but it does the job beautifully. Figure 5.1 shows why. Geometrically, if z is on the unit circle, $-z$ is at the point opposite it, 180° around the circle (because both the real and imaginary parts are negated). As z rotates counter-clockwise around the circle from 0 to π radians, $-z$ rotates in the same direction, from π to 2π radians. But those points also represent the negative frequencies from −π to 0, which for real signals are equivalent to the frequencies from π down to 0.

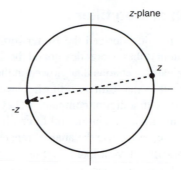

z-plane

Fig. 5.1 Replacing *z* by −*z* interchanges low and high frequencies.

We can also look at the transformation algebraically. Multiplying z by -1 is the same as multiplying by $e^{j\pi}$, so if we multiply a value of z on the unit circle at frequency ω by -1, we get

$$-e^{j\omega} \;=\; e^{j\pi}e^{j\omega} \;=\; e^{j(\omega+\pi)} \tag{5.1}$$

which is just another way of showing that π is added to each frequency. In a nutshell, we're just rotating the frequency circle 180°.

It's obvious, but worth emphasizing, that the transformation also multiplies the pole and zero locations by -1. Thus, the highpass version of the N-pole Butterworth filter has N zeros at $z = 1$ instead of at $z = -1$. Figure 5.2 illustrates the fact that the transformation shifts the Butterworth frequency response by the Nyquist frequency.

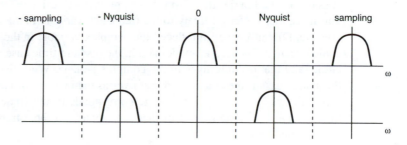

Fig. 5.2 Replacing *z* by −*z* in a transfer function shifts the frequency response by the Nyquist frequency, and thereby converts a lowpass filter, shown at the top, to a highpass filter, on the bottom.

We see that we can get highpass filters from a prototype lowpass filter. What about bandpass or bandstop designs? The same kind of trick works, except that the transformations are more complicated. You can learn more about it in more advanced books.

See, for example, the handbook edited by Mitra and Kaiser, or the book by Parks and Burrus, both cited in the Notes.

6 Connection with analog filters

The way I've just derived the Butterworth filter reverses history. It's as if I claimed that vacuum tubes were developed because people first had transistors, and then wanted gigantic versions that glowed in the dark.

The truth is that Butterworth filters were well known long before anyone dreamed of filtering with a digital computer. They were derived for analog filters. When people started thinking about digital filters, it was a natural idea to make use of the work already done, and to use the analog transfer functions to derive ones for digital filters.

Historically, it went like this. For analog filters, the lowest frequency is zero, as in the digital case; but there is no finite highest frequency. In other words, the range of frequencies is zero to infinity. We've seen that many times in connection with the Fourier transform (see Section 1 of Chapter 9, for example). To avoid confusion, from now on we'll use the symbol Ω for the analog frequency variable. (Up to his point we've been using the same ω for both analog and digital frequency, because there was no chance of confusing the two.)

We can now use the same reasoning as before to guess a good magnitude-squared frequency response for a lowpass filter. We want the frequency response to be one when $\Omega = 0$, and zero when $\Omega = \infty$. The form is even simpler than in the digital case; just try

$$\frac{1}{1 + \Omega^{2N}} \tag{6.1}$$

This has exactly the kind of lowpass behavior we want as Ω goes from zero to infinity. It has a cutoff frequency at $\Omega = 1$.

I now need to go back and reveal something I kept hidden from you in earlier discussions of the Fourier transform. If you review Table 1.1 in Chapter 11, for example, you'll see that I referred freely to the frequency variable in the analog case (which we now call Ω), but I never introduced the complex variable in the analog situation that is analogous to z. The frequency domain in the discrete-time case corresponds to the unit circle, and that ω-circle lives in the complex z-plane. Well, the corresponding axis in the analog case is the Ω-axis, and that lives as the imaginary axis in the complex plane called the s-plane. The point $s = j\Omega$ corresponds to the frequency of a continuous-time phasor, $e^{j\Omega t}$. The transform that corresponds to the z-transform is therefore the Fourier transform

$$X(\Omega) = \int_{-\infty}^{\infty} x(t) e^{-j\Omega t} dt \tag{6.2}$$

with $j\Omega$ replaced by the complex variable s:

$$X(s) = \int_{-\infty}^{\infty} x(t) e^{-st} dt \tag{6.3}$$

Thus, the usual frequency content $X(\Omega)$ is equal to $\mathcal{X}(s)$ when $s = j\Omega$, which is why we distinguish between X and $\mathcal{X}$. If you've taken electrical engineering courses, Eq. 6.3 should be very familiar — it's called the *Laplace transform*.[†]

As I've stressed all along, especially in Chapters 9 and 11, the mechanics of the Fourier (and hence Laplace) transform are beautifully analogous to those of the z-transform (although historians might say it is the other way around). As always, we can take advantage of the analogies to gain intuition. Most important to understand right now is that designers of analog filters can realize transfer functions that are ratios of polynomials in s, just as designers of feedback digital filters can realize transfer functions that are ratios of polynomials in z. Both kinds of filters are characterized by their poles and zeros. The only essential mathematical difference is in the frequency axes, the imaginary axis in the s-plane versus the unit circle in the z-plane. And this difference is reflected in the rest of those complex planes. The result is that analog filter poles must be in the left-half s-plane to correspond to the stable behavior of exponentially decaying time functions.

Returning to the connection between analog and digital filter design, compare Eq. 6.1 with Eq. 3.2:

$$\mathcal{B}(z)\,\mathcal{B}(z^{-1}) = \frac{1}{1 + (-F^2)^N} \tag{6.4}$$

The form is the same, and our function $-F^2$ corresponds to Ω^2, so F corresponds to $j\Omega$. In other words, if we think of F as a complex variable — and we did when we found the poles of the Butterworth digital filter — then F corresponds perfectly to the Laplace variable s.

The real history, then, is that analog Butterworth filters were designed in 1930 (see the Notes) starting with the function

$$\frac{1}{1 + (-s^2)^N} \tag{6.5}$$

(which becomes Eq. 6.1 when $s = j\Omega$) and the digital version was obtained about 30 years later by substituting

$$s = \frac{z-1}{z+1} \tag{6.6}$$

(which is just our function F as in Eq. 3.6).

The transformation in Eq. 6.6 is exactly what is needed to translate transfer functions for analog filters to transfer functions for digital filters. To begin with, the unit circle in the z-plane corresponds to the imaginary axis in the s-plane. Furthermore,

---

[†] There is a technical difference between the Laplace transform evaluated on the Ω-axis and the Fourier transform. The distinction is important in more advanced work. See A. Papoulis, *The Fourier Integral and its Applications,* McGraw-Hill, New York, N.Y., 1962.

the inside of the unit circle in the z-plane corresponds to the left-half s-plane. Digital filters with poles only inside the unit z-circle correspond under this transformation to analog filters with poles only in the left-half s-plane, so stable digital filters correspond to stable analog filters (see Fig. 6.1). The transformation that warps the Ω-axis so that $\Omega = 1$ in the analog world corresponds to $\omega = \omega_c$ in the digital world is easy enough to guess by analogy from Eqs. 4.3 and 4.5:

$$s = \frac{1}{\tan(\omega_c/2)} \cdot \frac{z-1}{z+1} \quad \text{and the inverse} \quad z = \frac{1+s\cdot\tan(\omega_c/2)}{1-s\cdot\tan(\omega_c/2)} \tag{6.7}$$

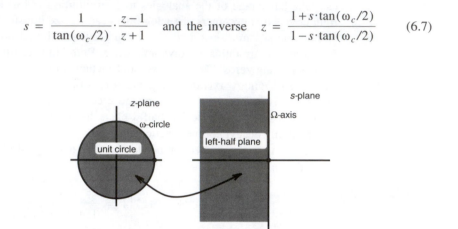

Fig. 6.1 Illustrating the transform $s = (z-1)/(z+1)$ and its inverse $z = (1+s)/(1-s)$. The key properties are the matching of unit ω-circle in the z-plane with the imaginary Ω-axis in the s-plane, and the inside of the unit z-circle with the left-half s-plane. The matching zero-frequency points are indicated by black dots, and occur at $z = 1$ and $s = 0$.

This transformation is called a *bilinear transformation*, and allows us to convert any analog filter transfer function to a digital one and vice versa. Intuitively, it compresses the infinitely long Ω-axis nonlinearly, and then very neatly wraps it once around the ω-circle. The infinitely high analog frequency, which can be thought of as a point at infinity in the s-plane, is mapped to the point $z = -1$ in the z-plane, the point corresponding to the Nyquist frequency. (Let $s \rightarrow \infty$ in Eq. 6.7.) The zero analog frequency, at the origin $s = 0$ in the s-plane, is mapped to the point $z = 1$. (Let $s = 0$ in Eq. 6.7.) The frequency response of the filter is squeezed or expanded like an accordion, but the approximation to an ideal lowpass or highpass shape is preserved. To see exactly how the two frequency axes are related, let $s = j\Omega$ and $z = e^{j\omega}$ in Eq. 6.7, and repeat the standard maneuver used to get from Eq. 2.6 to 2.7, yielding:

$$\Omega = \frac{\tan(\omega/2)}{\tan(\omega_c/2)} \tag{6.8}$$

This is the mathematical expression of wrapping the ω-circle to the Ω-axis, with the point $\omega = \omega_c$ corresponding to $\Omega = 1$, the cutoff frequency of the filter response we started with in Eq. 6.1.

To summarize the properties of the bilinear transformation,

s	$\Longleftrightarrow$	z
analog	$\Longleftrightarrow$	digital
imaginary axis	$\Longleftrightarrow$	unit circle
left-half plane	$\Longleftrightarrow$	inside unit circle
$\Omega = 0\,\text{Hz}$	$\Longleftrightarrow$	$\omega = 0$ radians per sample
$\Omega = \infty\,\text{Hz}$	$\Longleftrightarrow$	$\omega = \pi$ radians per sample

Here's a warning: The bilinear transformation relates the analog and digital worlds. But don't make the mistake of thinking that it corresponds in some way to the sampling process. It definitely doesn't (see Problem 12).

The 1930s was the golden age of analog filter design, spurred on by the needs of the telephone company (note the singular). The 1960s was similarly the golden age of digital filter design (and other things), but the hard work was already done — because the transformation in Eq. 6.7 was figured out. The other important analog filters used as prototypes for digital filters are the Chebyshev and elliptic filters, which achieve optimal approximations to a desired frequency response in the same mini-max sense that METEOR does for feedforward filters. We saw an example of an elliptic filter in Section 8 of Chapter 5. The frequency response of another is shown in Fig. 6.2.

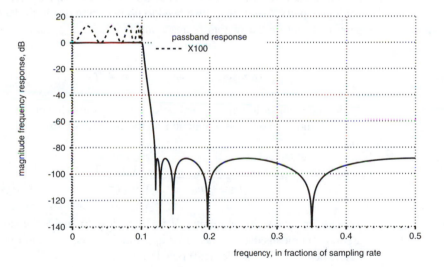

Fig. 6.2 Frequency response of a typical elliptic lowpass filter, illustrating the minimax character of its approximation to the ideal response, in contrast to the flat character of the Butterworth filter. The design specifications were to achieve at least 80 dB rejection in a stopband from 0.12 to 0.5 times the sampling rate, and rise no more than 0.2 dB above unity (which is 0 dB) in a passband from 0 to 0.1 times the sampling rate. The result is achieved with 10 poles and zeros.

Chebyshev and elliptic filters are designed the same way as Butterworth filters, except the expression $(-s^2)^{2N}$ in the denominator of Eq. 6.5, the squared magnitude of the frequency response, is replaced by more complicated polynomials that have special properties. This is a subject for more advanced books.

To summarize: We've derived the transfer function of the very useful Butterworth lowpass digital filter, seen how to move its cutoff frequency anywhere we want, and seen how to convert it to a highpass filter. We did it from scratch, but historically it was done by building on the earlier development of analog filters.

7 Implementation

The final step in designing any digital filter is deciding how it will be implemented — actually put into action. Let's implement the Butterworth transfer function in Eq. 4.7, which is in terms of the poles z_i, which in turn are given by Eq. 4.6.

First we need to straighten out a small complication. The poles are complex numbers, and we want our digital filter update equations to use only real numbers. We really don't want to store and process digital signals as complex numbers in the intermediate steps of filtering, because it would entail a lot of extra bookkeeping and require two memory locations for every sample. Fortunately, there's a simple way to put the transfer function in terms of real numbers. Complex poles occur in complex-conjugate pairs; therefore, all we need to do is form a quadratic factor from each such pair. Just replace the factors representing the poles at z_i and its conjugate z_i^* by the factor

$$(z - z_i)(z - z_i^*) = z^2 - 2\,\mathcal{Real}\,\{z_i\}z + |z_i|^2 \tag{7.1}$$

It's convenient to pair the real poles into quadratic factors as well, just so we can think of the whole transfer function as the product of such second-order factors, as follows:

$$\mathcal{B}(z) = \frac{A(z + 1)^N}{(z^2 + c_0 z + d_0) \cdots (z^2 + c_{N/2-1}z + d_{N/2-1})} \tag{7.2}$$

Remember that we're assuming N is even, just to make the notation simple. The assumption means that we have $N/2$ quadratic factors, and don't have to deal with the case when there is a single (real) pole left over.

Having $\mathcal{B}(z)$ in factored form makes it easy to come up with a simple and effective way to implement the filter. First, rewrite the transfer function in Eq. 7.2 by multiplying top and bottom by z^{-N}, in order to put the zero and pole factors in terms of the delay operator z^{-1}:

$$\mathcal{B}(z) = \frac{A(1 + z^{-1})^N}{(1 + c_0 z^{-1} + d_0 z^{-2}) \cdots (1 + c_{N/2-1}z^{-1} + d_{N/2-1}z^{-2})} \tag{7.3}$$

The real numbers c_i and d_i are easy to get in terms of the poles z_i, as indicated in Eq. 7.1. By comparing coefficients,

$$c_i = -2 \, \mathcal{R}eal\{z_i\} \tag{7.4}$$

and

$$d_i = |z_i|^2 \tag{7.5}$$

Finally, write the transfer function as a chain of subfilters, each having two poles and two zeros (remember that we're assuming N is even):

$$\mathcal{B}(z) = A \frac{(1 + z^{-1})^2}{(1 + c_0 z^{-1} + d_0 z^{-2})} \cdots \frac{(1 + z^{-1})^2}{(1 + c_{N/2-1} z^{-1} + d_{N/2-1} z^{-2})} \tag{7.6}$$

This form can be interpreted as a row of $N/2$ filters, the output of each filter feeding the next in line — in standard terminology a *cascade* of *second-order* sections. (See Fig. 7.1.)

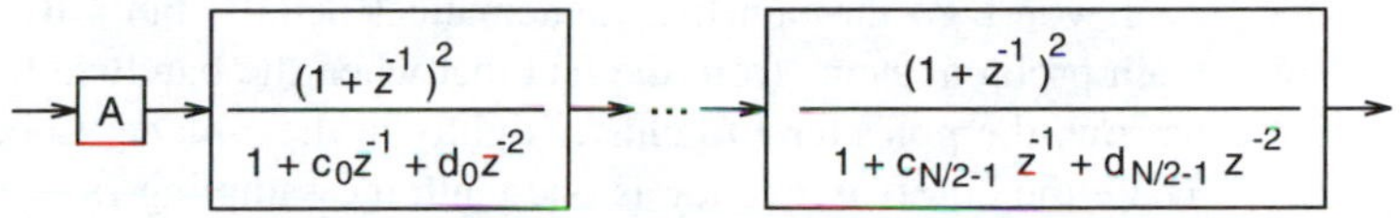

Fig. 7.1 Cascade implementation of the half-band Butterworth filter. This is the case when the number of poles is even, so there are $N/2$ subfilters.

We already know how to implement each second-order section. We could implement the numerator as two successive feedforward filters, each with transfer function $1 + z^{-1}$, but instead we'll expand the numerator as follows:

$$(1 + z^{-1})^2 = 1 + 2z^{-1} + z^{-2} \tag{7.7}$$

That way each stage (say stage i) can be described by the one simple update equation:

$$y_t = x_t + 2x_{t-1} + x_{t-2} - c_i y_{t-1} - d_i y_{t-2} \tag{7.8}$$

where x_t and y_t are the input and output signals of the ith stage.

The cascade form using second-order sections works for any feedback filter, not just Butterworth filters, and is widely used and very practical. Just factor the numerator and denominator of whatever transfer function you have, and assign pairs of poles and zeros to the sections. Each complex-conjugate pair of complex poles contributes a denominator of a section; and similarly for zeros in numerators. The real poles and zeros can be paired any way we want, although there is a subtle reason why some pairings and orderings of sections are better than others (see Problem 11).

8 A trap

At some point in your life, you may be tempted to implement a feedback filter by using the coefficients of the denominator directly, instead of using the poles two at a time, as I just recommended. If you're given the transfer function

$$\frac{a_0 + a_1 z^{-1} + \cdots + a_{m-1} z^{-(m-1)}}{1 + b_1 z^{-1} + \cdots + b_n z^{-n}} \tag{8.1}$$

why not implement it in the most straightforward way, using the following update equation

$$y_t = a_0 x_t + a_1 x_{t-1} + \cdots + a_{m-1} x_{t-(m-1)}$$
$$- b_1 y_{t-1} - \cdots - b_n y_{t-n} \tag{8.2}$$

where, as usual, x_t and y_t are the input and output signals? The answer may surprise you: Doing this can often lead to complete catastrophe; the filter may not work at all. The explanation was originally given by James Kaiser in 1965, and it was a key observation in the early stages of making digital filters work (see the Notes).

I won't go through the mathematical details, but I'll sketch the basic idea. The main problem stems from the fact that when the bandwidth of a digital filter gets very narrow, the poles tend to cluster tightly in the z-plane. Look at Fig. 4.1, for example. There the cutoff frequency is one-tenth the sampling rate, and the poles have already moved off the imaginary axis (where they were when the cutoff was one-quarter the sampling rate) and are headed for the point $z = 1$. As the bandwidth gets narrower and narrower the poles will collect very tightly around that point. Intuitively, if the frequency response has a very narrow passband near some frequency, the poles must be placed close to the spot on the frequency circle where all the action is.

Now it's well known in numerical analysis that when the roots of a polynomial equation are clustered around a point, the positions of the roots themselves are extremely sensitive to tiny errors in the coefficients of the polynomial. Even in reasonable digital filter designs, the pole positions can be so sensitive to perturbations in the denominator coefficients that the usual 32 or 64 bits used to store them are not enough to fix the poles where we want them with sufficient accuracy. In fact, they can easily go shooting off outside the unit circle, and the resulting digital filter will be unstable and worthless.

9 Feedback versus feedforward

Which should you use for a given filtering job, feedback or feedforward filters? I raised this question at the end of Chapter 5, and had a bit to say about it then. I'll add more now, but the general issue is complicated.

Soon after design methods were developed for both kinds of filters, Rabiner et al. (see the Notes) carried out a detailed study using lowpass filtering as the representative task, and elliptic filters as the representative feedback design. The general conclusion was that if you count the number of arithmetic operations, elliptic filters are in many cases much more efficient than feedforward filters, *in terms of operation count*. Generally speaking, feedback filters give you more ''bang for the buck'' than feedforward filters. As we've seen, poles are a lot more influential in the complex plane than zeros.

But there's more to the story than simply counting computer instructions. You may not be very concerned about how long it takes to carry out the filtering operation, within reason — it all depends on how much filtering you're planning to do. If raw efficiency is not important, feedforward filters have some real advantages. We learned in Chapter 5 that feedforward filters are easy to design with exactly linear phase, which amounts to having no phase distortion. And we saw in Chapter 12 that it's much easier to design feedforward filters with completely arbitrary magnitude response.

A third important advantage of feedforward filters has to do with what are usually called *finite wordlength effects* — deviations from theoretical operation caused by the fact that arithmetic is carried out with a finite number of bits. These effects include roundoff error, as well as the problem of representing the filter coefficients themselves in finite computer words, which we encountered in Section 8. Generally speaking, feedforward filters are much better behaved in this department, because their transfer functions are linear in their coefficients and because they have no feedback.

In this chapter we've just had time to touch on the basic ideas of designing closed-form feedback filters. Together with the feedforward filters you can design with programs like METEOR, they provide all you will need in many practical situations.

We've now introduced the main ideas and tools of digital signal processing. You should be able to use the commonly available programs, like those for the FFT, filter design, and filtering, with good sense and understanding; and you should also be well equipped for more advanced study. We'll finish this book with a sampling of some applications that further illustrate these ideas, and that are also important in audio and computer music.

Notes

The general design problem for feedback filters is difficult, but not impossible. Practical design packages today use iterative optimization algorithms, still similar in approach to the early, influential paper:

A. G. Deczky, ''Synthesis of Recursive Digital Filters Using the Minimum *p*-Error criterion,'' *IEEE Trans. Audio and Electroacoustics*, vol. AU-20, no. 4, pp. 257–263, October 1972.

Butterworth achieved immortality with the following elegant six-page paper:

S. Butterworth, ''On the Theory of Filter Amplifiers,'' *Wireless Engineer*, vol. 7, pp. 536–541, 1930.

Not only did he think of putting the poles equally spaced on a circle in the *s*-plane (Eq. 6.5), but he knew how to get them there with a soldering iron:

"The writer has constructed filter units in which the resistances and inductances are wound round a cylinder of length 3 in. and diameter 1¼ in., while the necessary condensers are contained within the core of the cylinder."

The wonderful properties of the bilinear transformation have been well known to mathematicians and scientists for at least a couple hundred years. I used it to design digital filters and to relate the analog and digital worlds in my thesis, which of course I can't resist referencing:

K. Steiglitz, "The General Theory of Digital Filters with Applications to Spectral Analysis," Eng. Sc.D. Dissertation, New York University, New York, N.Y., May 1963.

Around the same time, independently, just across the Hudson River, James Kaiser was having similar thoughts, and he describes and applies the idea in

J. F. Kaiser, "Design Methods for Sampled Data Filters," *Proc. First Allerton Conf. on Circuit and System Theory*, Urbana, Ill., pp. 221–236, Nov. 1963.

The paper mentioned in Section 9 comparing feedforward (FIR) and feedback (IIR) filters is

L. R. Rabiner, J. F. Kaiser, O. Herrmann, and M. T. Dolan, "Some Comparisons Between FIR and IIR Digital Filters," *Bell System Technical Journal*, vol. 53, pp. 305–331, Feb. 1974.

If you want a good snapshot of the way things looked at the beginning of the 1970s, when theory and technology were coming together with explosive force, see the following collection of papers:

L. R. Rabiner and C. M. Rader (eds.), *Digital Signal Processing,* IEEE Press, New York, N.Y., 1972.

As in the case of feedforward filters, the following are rich sources for more information and references about digital filter design:

S. K. Mitra and J. F. Kaiser (eds.), *Handbook for Digital Signal Processing*, John Wiley, New York, N.Y., 1993.

T. W. Parks and C. S. Burrus, *Digital Filter Design*, John Wiley & Sons, New York, N. Y., 1987.

A. V. Oppenheim and R. W. Schafer, *Digital Signal Processing*, Prentice-Hall, Englewood Cliffs, N.J., 1975.

L. R. Rabiner and B. Gold, *Theory and Application of Digital Signal Processing,* Prentice-Hall, Englewood Cliffs, N.J., 1975.

Problems

1. Suppose we vary any one coefficient of a feedforward filter while keeping all the others constant, and we plot the maximum value of the magnitude frequency response. In general, what will the curve look like? Suppose we plot instead the maximum value of the error between the actual frequency response and some fixed, prespecified frequency response. What will the general shape of the curve be then?

2. Put the transfer function in Eq. 2.1 in a more familiar form, showing it really represents a feedforward filter.

3. I said that when we split the function $\mathcal{B}(z)\,\mathcal{B}(z^{-1})$ in Eq. 2.5, we would choose the zeros and poles inside the unit circle to form $\mathcal{B}(z)$. Suppose that instead we choose some zeros outside the circle. What effect will that have on the resulting filter transfer function?

4. Find explicit expressions for the gain constants A in Eqs. 3.10 and 4.7 in terms of the poles of those transfer functions.

5. Suppose we want to design a half-band Butterworth filter with prespecified frequency response value at a given right edge of the passband, say at $\omega = \omega_1$; and also a prespecified response value at a given left edge of the stopband, say at $\omega = \omega_2$. Assume, of course, that $\omega_1 < \omega_2$. Derive the design equations that determine the choice of the order N. Hint: Use Eq. 2.8.

6. Verify that the $2N$ poles in the F-plane when developing the N-pole Butterworth filter are as shown in Fig. 3.1 and Eq. 3.5.

7. Revise the relevant equations in Sections 3 and 7, as well as Fig. 7.1, for the case when N is odd. Where is the real pole in the half-band case? Where is it when the cut-off frequency is ω_c?

8. What changes should you make to the coefficients c_i and d_i in the transfer function in Eq. 7.6 to convert a lowpass Butterworth filter to highpass?

9. We mentioned two ways to implement a feedback filter: the cascade form in Fig. 7.1, and the expanded form in Eq. 8.2, called the *direct form*. Determine how many additions and multiplications are required for each. Which is more efficient in this respect?

10. Butterworth filters are sometime called ''maximally flat.'' Explain qualitatively, then verify mathematically.

11. As mentioned in Section 7, some pairings and orderings of the sections in a cascade implementation of a feedback filter may be better than others. Why?

12. The bilinear transform can be used to relate the frequency transforms of digital and analog signals, by using $X((z-1)/(z+1))$ in place of the Laplace transform $X(s)$, for example. As noted in Section 6, however, this transformation does not correspond to the sampling process. Find a specific counterexample to the notion, using the unit step signal. (Use the definition in Eq. 6.3 to find the Laplace transform $X(s)$ of the unit

step signal.) How are the transforms of an analog signal and its sampled version actually related?

13. Prove that the poles of the lowpass Butterworth filter with general cutoff frequency, like the one used as an example in Figs. 4.1 and 4.2, lie on a circle in the z-plane.

14. Can you tell from the frequency response shown in Fig. 6.2 how many poles and zeros the corresponding elliptic filter has? Explain.

15. What transformation in the s-plane corresponds to the lowpass-to-highpass transformation $z = -z$? Here's a hint from Butterworth, who writes in his 1930 paper referenced in the Notes: "In the low pass system let the inductances be replaced by capacities and the capacities by inductances."

CHAPTER

14

Audio and Musical Applications

1 The CD player

We'll begin our small tour of audio applications with the compact disc (CD) digital audio system, a wonder of our age. The CD is conceptually a very simple storage system: the audio signal is sampled and converted from analog to digital form, and the bits are stored on the surface of a small platter. When we want to listen to them, we just spin the disc, read off the bits, and convert the digital signal back to analog form. This should seem like a simple and natural idea by now — but its practical realization has changed the world forever. As I mentioned at the very beginning of this book, digital audio makes it possible, for the first time, to grasp music securely in our hands.

The technology of the CD audio system we use today was developed in the mid- to late 1970s and depends critically on the very dense storage of bits on an optically scanned disc.[†] The spiral track on the standard 12 cm disc is about three miles long, and the tracks are 1.6 μm apart. (Recall that a μm, also called a micron, is a millionth of a meter, or about 1/25,000th of an inch.) The track itself is 0.6 μm wide, and the spot of laser light used to scan it is 1 μm in diameter. The bits are actually stored in the form of depressed regions along the track (''pits'') and nondepressed regions at surface level (''lands''). A great advantage of this arrangement is that nothing but light ever touches the storage area of the disc. In theory it can last forever.

The CD audio system is conceptually simple, but engineers have developed some clever twists and turns, either to improve performance without raising cost, or to lower cost. The basic digital signal uses 16 bits per sample at a sampling rate of 44.1 kHz, so the two tracks for stereo require a basic bit rate of $2 \times 16 \times 44,100 = 1.41$ Mbit/sec. It may come as a surprise that the actual data rate onto and off the disc is about three

---

[†] The Philips and Sony companies played central roles in the development. See the Notes for reference to a write-up by Philips staff.

times as fast. The extra bits are used for several purposes, including redundancy for detecting and muting errors, synchronization, and the embedding of handy tidbits like the name of the piece, its duration, and its track number. The interweaving of the extra, nondata bits is Byzantine in its complexity, and its description belongs more properly in a course on coding theory.

In this section I'll focus on the process of retrieving the bits from the disc and converting them to the analog signal we hear. This is digital signal processing at its best, and we're now well equipped to understand it.

Recall that in Section 8 of Chapter 11 we described what we called *oversampling*, where we sampled an analog signal at a higher rate than necessary, so that we could make the anti-aliasing analog prefiltering easy. This had the effect of transferring the hard part of the filtering to the digital domain, where, as it turns out, we can do a much better job, much more easily. A reduction in sampling rate (sometimes called *down-conversion*) takes place after analog-to-digital (a-to-d) conversion.

We can use the same general idea for digital-to-analog (d-to-a) conversion. If we did the d-to-a conversion in the CD system at the true data rate of 44.1 kHz, we would generate images above the Nyquist frequency of 22.05 kHz, and we'd need an analog filter with a sharp cutoff to suppress them. The desired audio band in this case extends up to 20 kHz, so the frequency response of the analog filter would have to make the transition from passband to stopband in the band of frequencies between 20 kHz and 22.05 kHz — not a lot of room. An analog filter with such a narrow transition band, in this range of frequencies, would be expensive. The alternative is to do the a-to-d conversion at a higher frequency, and do most of the filtering work in the digital domain.

We have a terminology problem here. People use the term ''oversampling'' in both the a-to-d and d-to-a cases — but, while the basic trick is similar, the methods are quite distinct. I'll distinguish the two with the terms ''oversampled a-to-d conversion'' and ''oversampled d-to-a conversion.'' It's the latter we're discussing here.

The Philips system increases the sampling rate by a factor of four before d-to-a conversion. This process, called ''up-conversion,'' can be thought of as interpolating reasonable values between the actual signal samples. There is a beautiful way to do this with digital filters that is the opposite of the down-conversion we did after oversampled a-to-d conversion in Section 8 of Chapter 11.

We start the interpolation process by doing the seemingly silly thing of inserting zeros where the missing, interpolated samples should go. What is the spectrum of the new signal? Well, the z-transform is just a power series, and the only nonzero terms occur now at points where the sample number is a multiple of four. That is, the original samples occur at new sample points 0, 4, 8, and so on. If the original z-transform is $X(z)$, the new transform is therefore just $X(z^4)$; z is replaced everywhere by z^4. (I asked you to derive this in Problem 2 of Chapter 9.)

If we now trace the frequency variable ω from 0 to π, the new complex frequency argument of X, z^4, will move on the unit circle from the angle 0 to 4π, which means that the segment of the original spectrum between 0 and π will be traced out four times. This is illustrated in Fig. 1.1(a) and (b). The net effect is to move the Nyquist and sampling frequencies up by a factor of four.

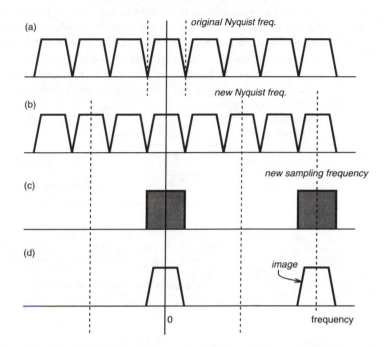

Fig. 1.1 Oversampled d-to-a conversion avoids expensive analog postfiltering. The figure shows signal spectra during the digital processing used to increase the sampling rate before conversion: (a) original digital signal; (b) interleaving three zero-valued samples between successive signal samples moves the Nyquist frequency up by a factor of four; (c) the desired quarter-band digital filter frequency response to prepare for d-to-a conversion at the higher rate; (d) the signal spectrum after digital filtering. At this point the signal is ready for d-to-a conversion at four times the original rate.

If we just convert at four times the original rate (176.4 kHz) we gain nothing; there will be images above the original Nyquist frequency, where they would have been anyway. The point is that now we can remove them digitally. Figure 1.1(c) shows the desired digital filter frequency response, which is quarter-band lowpass, and Fig. 1.1(d) shows the final spectrum of the digital signal after the filtering operation. The d-to-a conversion is actually followed by a zero-order hold, and some of the image shown centered at four times the original sampling rate will get through (see Problem 3).

That image is centered at 176.4 kHz, way beyond the range of hearing, and it doesn't take much of an analog filter to do a good job removing it (but see Problem 4). The transition band available to the analog postfilter is now enormous compared to what it was without the up-conversion and digital filtering process. To put it simply, we have bludgeoned the problem with blindingly fast digital processing and made the work remaining in the analog domain simple and cheap.

Looking at the time waveforms makes it obvious why the process of oversampled d-to-a conversion makes life easier in the analog, post-conversion world. Figure 1.2 shows the stages, starting with the original samples. Removing the images above the original Nyquist frequency is done by the quarter-band digital filter, and amounts to interpolating smoothly between the original samples. It's not hard to believe that the converted signal in Part (d) of the figure is a lot easier to clean up with a post-conversion analog smoothing filter than the much choppier signal in Part (a).

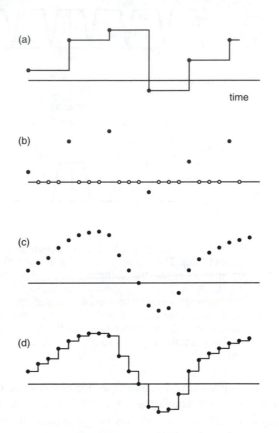

Fig. 1.2 Up-sampling in the time domain, where it's easy to interpret as interpolation. (a) the original signal samples and what a zero-order hold would do to them; (b) interleaving zeros; (c) the smoothed result after quarter-band digital filtering; (d) the final output of a zero-order hold when d-to-a conversion takes place at four times the original rate.

The quarter-band filter used in the Philips system is a feedforward filter with 96 coefficients. An important advantage of a feedforward over a feedback filter is its linear phase. Another is the short word-length possible. The filter used in the Philips system uses only 12-bit coefficients, and the multiplications by the 16-bit samples produce 28-bit outputs, which are of course rounded off for conversion. A third advantage results from the fact that of the 96 multiplications apparently required for each output sample, three-quarters are multiplications by zero, so only 24 nontrivial multiplications per output sample are actually needed.

Still another advantage of feedforward filters stems from our being able to design them to satisfy quite arbitrary frequency domain specifications. In fact, the specifications for the quarter-band filter are unusual in a couple of respects. For one thing, the stopband rejection is chosen so that the trench is deeper in the region immediately following the original Nyquist frequency than the rest of the stopband. (Why? See Problem 5.) For another, the passband is shaped to compensate for the "droop" caused by the zero-order hold, as well as for the ripple caused by the simple analog postfilter (which has three poles). The paper by Goedhart et al. referenced in the Notes shows the frequency response of the actual filter used in the Philips CD player, and I mimicked the design using the METEOR program discussed in Chapter 12. Figure 1.3 shows my resulting frequency response. I chose constraints that increased linearly on a dB scale, with ±3 dB ripple in the passband, and rejection in the stopband going from −50 dB at 24 kHz to −30 dB at the Nyquist frequency of 88.2 kHz. I don't know what design method Goedhart et al. used, but my filter response is quite similar, although they show the result after rounding the coefficients to 12 bits.

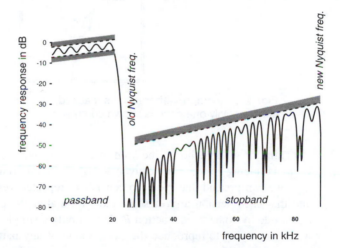

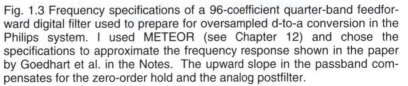

Fig. 1.3 Frequency specifications of a 96-coefficient quarter-band feedforward digital filter used to prepare for oversampled d-to-a conversion in the Philips system. I used METEOR (see Chapter 12) and chose the specifications to approximate the frequency response shown in the paper by Goedhart et al. in the Notes. The upward slope in the passband compensates for the zero-order hold and the analog postfilter.

We see now that designing feedforward filters to meet arbitrary specifications is more than a rainy-afternoon amusement (although it is that too). The resulting little pieces of technology find their way into one of the most common artifacts of the late twentieth century.

2   Reverb

Reverb, as it's always called by rock guitarists and other professional sound makers, is another important application of digital signal processing, and of digital filtering in particular. Bare computer music, for example, has a tendency to sound dry, simply because it is created in what amounts to an anechoic chamber. Composers often use reverb to liven it up. And without reverb the drive-time news wouldn't sound as if the radio studio is a vast crypt.

Reverberation is the general term for what happens to a sound when it makes its way from the source, be it mechanical or electronic, to our ears. It bounces off the walls and other parts of rooms and concert halls and gets mixed up with its echoes, as shown in Fig. 2.1.

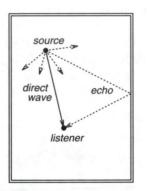

Fig. 2.1 Room reverberation is caused by the superposition of many echoes. Only one echo is shown completely.

Think about an echo — the kind of echo you hear in a canyon, where the sound keeps bouncing between the canyon walls. To get a crude mathematical approximation, we can pretend that the echo consists of repeated versions of the original sound, and that each version arrives some fixed time after the previous version, reduced in amplitude by some fixed fraction R. This is only a simple starting point, but, after all, we're not trying to reproduce the exact sound of any particular environment. All we want is a transformation that gives us the general effect of an echo. If the original sound is represented by the digital signal x_t, the echoed version y_t is

$$y_t = x_{t-L} + Rx_{t-2L} + R^2 x_{t-3L} + \cdots \tag{2.1}$$

Ideally this goes on forever, but we know that real echoes diminish in amplitude until they merge with the background noise.

Taking the z-transform of both sides of Eq. 2.1 gives us the following relationship between the transforms of x and y:

$$\mathcal{Y}(z) = X(z)\left[z^{-L} + Rz^{-2L} + R^2 z^{-3L} + \cdots\right] \tag{2.2}$$

Factor out a delay z^{-L}, and what is left is just a geometric series with the ratio Rz^{-L}

between successive terms. The ratio of transforms is therefore the familiar-looking closed form:

$$\frac{\mathscr{Y}(z)}{X(z)} = \frac{z^{-L}}{1 - Rz^{-L}} \qquad (2.3)$$

Except for the initial delay of L samples, this is the transfer function of a comb filter, which we studied in Chapter 6. The corresponding signal flowgraph is shown in Fig. 2.2. The only difference is that now we're thinking about round-trip delays that correspond to the distances between walls in a room or concert hall, whereas in Chapter 6 we were thinking about waves bouncing around a tube or string the size of a musical instrument.

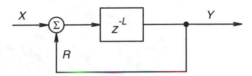

Fig. 2.2 A simple comb filter. The integer L is the loop delay in samples, R is a constant loop multiplier, and X and Y are the input and output signals, respectively.

Let's take a look at the resonant frequencies that result when we choose the delay in the comb filter to correspond to a hall of moderate size, say with a round-trip echo distance of 20m. The speed of sound is about 345 m/sec. (This easily remembered number corresponds to a nice warm room; see Benade's book referenced in the Notes to Chapter 3.) A typical round-trip time is therefore about 60 msec. By way of contrast, a column of air in a flute that is 1/3 m long, operating as a pipe open at both ends, will have a round-trip time 1/60th of this, or about one msec. (The mouth hole and first open tone hole approximate a tube open at both ends.) The corresponding lowest resonant frequency of an open pipe (recall Chapter 6) is the reciprocal of the round-trip time, or about 17 Hz for the hall, and 1 kHz for the flute.

Of course, one simple comb filter will not do the job we want. A typical room has many reflecting surfaces, and it's necessary to combine many filters to get anything like a realistic-sounding reverb. The best reverb filters have evolved through trial and error, guided by physical measurements of concert halls and by inspired intuition. I'll describe one of the best results, from a well-known article by James Moorer (see the Notes).

Tailoring a reverb filter to duplicate the sound of a particular room or concert hall is still more an art than a science. (The same may be said of designing concert halls themselves.) But the guiding principle is to avoid regularity. The usual recommendation is to make the round-trip delays in the combs, when expressed in number of samples, mutually prime. This tends to encourage the echoes to blend nicely. As Moorer puts it, it "reduces the effect of many peaks piling up on the same sample, thus leading to a more dense and uniform decay."

Moorer suggests the configuration shown in Fig. 2.3, consisting of six different comb filters in parallel (their outputs added), followed by an allpass filter. This allpass filter has the effect of further mixing up the phase of the signal without changing the magnitude of the frequency content. The path at the bottom with constant gain K represents the direct wave, the transmission of sound to the listener without reflections. Moorer also proposes refining each comb by putting a lowpass filter in the delay loop, as shown in Fig 2.4. The purpose of this is to take into account the fact that the absorption of sound depends on frequency, with high frequencies being absorbed more. He uses single-pole lowpass filters, as indicated in Fig. 2.4, as an approximation that captures the main effect without introducing too much complexity in the overall reverb filter.

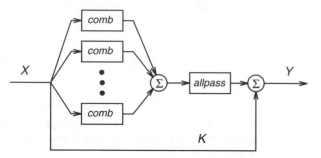

Fig. 2.3 A popular reverb filter structure, proposed by J. A. Moorer.

Suppose we've decided to use the lowpass-comb/allpass structure shown in Fig. 2.3. We still have many choices to make: the loop delays L, gain constants R, and pole positions g of the six lowpass combs; the direct-wave factor K; and the allpass parameters. Moorer discusses these choices on physical grounds, and gives the results of choosing the six g parameters to fit physical measurements of room acoustics in a least-square sense. Before I give some specific numbers in case you want to implement the reverb filter recommended by Moorer, I need to clear up a couple of details.

First, the lowpass comb filter shown in Fig. 2.4 may be unstable if the parameters g and R are chosen injudiciously. To get some intuition about how this can happen, we'll look at the transfer function of the lowpass comb. This is just the original comb transfer function, Eq. 2.3, with the lowpass filter transfer function inserted along with the R term:

$$\frac{\mathcal{Y}(z)}{X(z)} = \frac{z^{-L}}{1 - R\left[\dfrac{1}{1 - gz^{-1}}\right]z^{-L}} \qquad (2.4)$$

The term subtracted from one in the denominator, the so-called *loop gain*, represents the effect on a signal traveling once around the feedback loop.

Now the signal traveling around the loop — representing a bouncing sound wave — can be thought of as a sum of various frequency components. That's one of the main points of all our work on frequency transforms. Consider for the moment the zero-frequency component. The loop gain at zero frequency is the loop gain evaluated

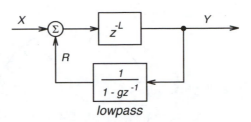

Fig. 2.4 A lowpass comb, a version of the comb component that takes into account frequency-dependent sound absorption.

at $z = 1$, the point corresponding to zero frequency on the unit circle in the z-plane. That loop gain turns out to be $R/(1 - g)$. If this is greater than one, the zero-frequency component would be amplified each time it traveled around the loop, and the accumulating signal would grow larger and larger without bound. In other words, the filter would be unstable, and we'd get nothing useful out. This couldn't happen in a real room, of course, unless the walls amplified echos — a frightening thought.

An analogous argument can be made at other frequencies, but then things are more complicated because the signals bounced back are complex, which is another way of saying there is generally phase shift. A better way to understand what's happening is to look at the poles of the overall lowpass comb as we vary the lowpass filter parameter g. Multiplying the top and bottom of the transfer function in Eq. 2.4 by $(1 - gz^{-1})$ and then by z^L, we get

$$\frac{\mathcal{Y}(z)}{X(z)} = \frac{z - g}{z(z^L - gz^{L-1} - R)} \tag{2.5}$$

which shows how to find the poles. Except for the pole at $z = 0$, they're roots of the polynomial equation

$$z^L - gz^{L-1} - R = 0 \tag{2.6}$$

We all have a program that factors polynomials, and it's perfectly feasible to factor this for lots of values of g.[†] Figure 2.5 shows a plot of the pole migration as g is increased from 0 to 10 in a lowpass comb with a loop delay of 20 samples and a loop multipler $R = 0.5$. As we know, the poles start out, when $g = 0$, at the 20 roots of 0.5, which is just the case of the ordinary comb filter. As g increases, some of the poles stray outside the unit circle. It turns out that the first one to leave in this situation is the pole on the positive real axis.

We can determine exactly when the pole on the positive real axis crosses the unit circle by examining the transfer function of the lowpass comb. The pole on the positive real axis must cross the unit circle at the point $z = 1$, and at that value of z the

---

[†] This is an example of a *root locus* plot, a very useful way to study the stability of a filter when a parameter is varied.

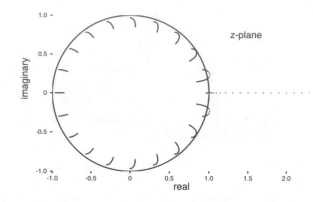

Fig. 2.5 The poles of the lowpass comb move when the pole-parameter g increases from zero. The case of a lowpass comb with a loop delay of 20 samples is shown here. When g exceeds 0.5, the pole on the positive real axis moves outside the unit circle, causing instability.

denominator is $1 - R/(1-g)$. Therefore the filter just reaches the point of instability when the loop gain at zero frequency, $R/(1-g)$, becomes unity. To keep the filter safely stable, we should choose this ratio to be less than one.

The second detail concerns the allpass filter that follows the parallel combs. The allpass filters we used in Section 6 of Chapter 6 to tune plucked-string filters had delay loops of only one sample, because we were interested in introducing delays on the order of a fraction of a sample. Now we want delays on the order of a few msec, to scramble echos. We can use the same transfer function, but with each unit delay replaced by a delay of m samples:

$$\mathcal{H}(z) = \frac{z^{-m} + a}{1 + az^{-m}} \qquad (2.7)$$

Problem 7 asks you to verify that this is still allpass, for any value of m.

If you want to try to implement a reverb, I'll tell you what Moorer recommends; it should give you a good starting point for experimentation. For the allpass filter in the lowpass-comb/allpass structure, Moorer recommends using a value of $a = 0.7$ and a delay of 6 msec. Convert the delay in seconds to a number of samples by multiplying by the sampling rate. He suggests the following set of delay values for the six lowpass-comb filters: 50, 56, 61, 68, 72, and 78 msec. The lowpass filter parameters g depend on the sampling rate, delay loop length, humidity, temperature, and pressure. Moorer provides tables showing the results of fitting the lowpass frequency response to published data describing high-frequency attenuation in air. The one-pole lowpass filter is so crude that these values should be taken only as guidelines. As an example, he suggests the values for the lowpass parameter g of 0.24, 0.26, 0.28, 0.29, 0.30, and 0.32, for the sampling rate of 25kHz. He determines the gain constant R for each comb by setting the zero-frequency loop gain $R/(1-g)$ to 0.83 for all the comb filters, which he says results in a reverberation time of about 2 sec. I've used all these parameters and a ratio of direct wave to allpass output of 9 to 1. The result provides the feeling of a very lively room.

Figure 2.6 shows the impulse response of the resulting reverb filter. After an initial dead space, there is a dense, complicated pattern of reflections, eventually decaying to zero — just what we want.

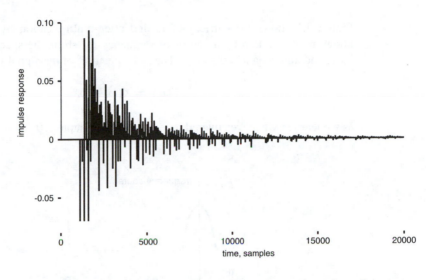

Fig. 2.6 Impulse response of a reverb filter put together along lines suggested by Moorer. The sampling rate is 22.05 kHz. The initial sample, representing the direct wave, is off-scale at 0.9.

3 AM and tunable filters

A remarkable thing about the frequency domain ideas we've developed is that we can use them in so many situations: for sound waves, radio waves, or light waves — not to mention electrical signals from the brain, heart, or even vibrational waves from the earth. The frequencies may be measured in cycles per century — for sunspots, say — or billions of cycles per second, for microwaves. We can take any idea that works in one area and apply it in another.

We'll now see that the principle behind AM (amplitude modulation) radio provides a very useful example of such technology transfer. AM radio is most familiar in the medium-wave broadcast band, which operates around 1 MHz, but the idea works perfectly well at audio frequencies. The mathematical principle is the heterodyne property of the Fourier transform, which we discussed in Chapter 10. I'll now explain briefly how AM radio works, and then show how heterodyning can be used to build bandpass digital filters that can be tuned very conveniently.

Radio waves can travel enormous distances, of course. Sounds waves cannot. The basic trick of radio communication is to allow audio waves to hitch a ride on radio waves. The process is called *modulating* the radio wave (the *carrier*) with the audio signal. The simplest way to do this is to multiply the audio signal by a radio signal at a single frequency — a phasor. Letting $x(t)$ stand for the real-valued audio signal, and using a radio signal of frequency ω_0, the product is the broadcast signal

$$y(t) = x(t)e^{j\omega_0 t} \tag{3.1}$$

Figure 3.1 shows the carrier before and after multiplication by the signal. As we know, multiplication by a phasor of frequency ω_0 shifts the spectrum of the original baseband audio signal up by ω_0. The new spectrum of the signal $y(t)$ is therefore

$$Y(\omega) = X(\omega - \omega_0) \tag{3.2}$$

as we learned when we covered heterodyning in Chapter 10.

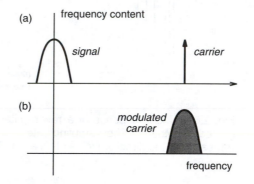

Fig. 3.1 AM modulation of a carrier by a baseband signal. Think of the modulation process as spreading the spectrum of the carrier.

It's very useful to interpret Eq. 3.2 and Fig. 3.1 as meaning that the baseband signal — the information-bearing signal — *spreads* the spectrum of the carrier. In fact, in AM it spreads the carrier spectrum to a width precisely equal to the bandwidth of the modulating signal, measured from negative to positive frequencies. That bandwidth is a precious commodity, and, as we've discussed before (Section 6 of Chapter 11), is directly related to the amount of information we can transmit. We'll have more to say about modulation when we discuss frequency modulation in the next section.

In practice, we multiply by a real-valued cosine or sine wave, which is the sum or difference of phasors at $\pm\omega_0$. The result would be two translated versions of the baseband spectrum, one at ω_0 and another at $-\omega_0$. That tends to clutter up the formulas, and in this chapter I'll stick to complex phasors and use complex signals. It's nice to be able to deal with just one frequency instead of with the positive and negative pair. The notation will be simpler and the ideas clearer. It turns out that when we're all done, though, we could have achieved the same overall effect with real arithmetic, multiplying by sines and cosines and combining the results (see Problem 10).

Suppose now that we want to build a bandpass digital filter with a center frequency that is very easy to vary. We might want to use such a tunable filter in a spectrum analyzer, for example, where we are interested in how much energy there is in many different bands. Think of this as a filter with a tuning knob on it — exactly what we would need for a receiver dealing with radio frequencies. We could just build many filters, each with a different center frequency, but that would be inconvenient, since we'd need to store a different set of coefficients for each center frequency of interest.

But here's a way to do the job with only one filter: move the signal frequencies instead of the frequency response of the filter. We just saw how to move the signal frequencies with heterodyning. If we shift the signal frequencies down by ω_0 and then apply a lowpass filter, as shown in Fig. 3.2, the frequencies selected by the filter will actually correspond to the vicinity of ω_0 in the original signal spectrum. We can then shift the frequencies selected by the lowpass filter back to their original position by heterodyning up by ω_0. This process produces the answer for positive frequencies and is illustrated in the top branch of Fig. 3.3. The analogous process produces the negative frequency part — shift up, lowpass filter, then shift back down. The two parts are added, as shown in Fig. 3.3.

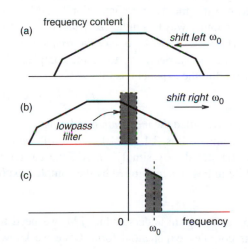

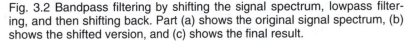

Fig. 3.2 Bandpass filtering by shifting the signal spectrum, lowpass filter-ing, and then shifting back. Part (a) shows the original signal spectrum, (b) shows the shifted version, and (c) shows the final result.

This trick is enormously useful, and, as I mentioned, allows us to use one fixed, carefully designed lowpass filter in a spectrum analyzer, for example, where we want to sweep over a wide range of frequencies. We'll return to this technique when we discuss vocoders, but first we'll visit the FM band.

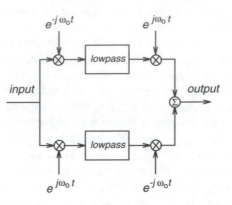

Fig. 3.3 Tunable bandpass filtering by heterodyning. The upper branch produces the positive-frequency part of the answer by shifting left, lowpass filtering, and shifting back. The lower branch makes the negative-frequency part in the analogous way.

4 FM synthesis

I'm going to take a moment to describe FM (frequency modulation) — not only because it's the natural companion to AM on your radio dial, but also because it's a very popular way to synthesize sound on electronic keyboard instruments.

The idea of FM is to vary the *frequency* of the carrier instead of its amplitude. If we take the carrier wave to be the complex phasor $e^{j\omega_0 t}$, a modulating signal $x(t)$ is inserted in the exponent as follows:

$$e^{j(\omega_0 t + Dx(t))} = e^{j\omega_0 t} e^{jDx(t)} \tag{4.1}$$

where D, called the *modulation index*, is a measure of how much we are modulating.[†]

Compare Eqs. 3.1 and 4.1, and you'll see that the difference between AM and FM can be stated very simply: In AM the carrier is multiplied by the signal $x(t)$ itself, while in FM it is multiplied by the complex exponentiation of the signal:

$$e^{jDx(t)} \tag{4.2}$$

To go further in understanding FM, we need to know the frequency components of this complex exponential form. Once we know that we're practically done, because the final modulated carrier will be just those components shifted up by the carrier frequency ω_0. With this in mind, we can forget about the carrier frequency; what matters is the spectrum of the signal in Eq. 4.2.

Begin with the simple case when the signal $x(t)$ — the information that will be impressed on the carrier — is a simple sine wave, say $\sin(\omega t)$. Notice that this $x(t)$ is periodic, with frequency ω radians per sec, and hence so is its complex exponentiation in Eq. 4.2. We can therefore expand it in a Fourier series, which will tell us exactly

---

[†] Strictly speaking, this equation shows the modulation of the phase angle, not the frequency, of the carrier. The difference needn't bother us here.

what frequency components are there, and in exactly what proportion. The fundamental frequency being ω, the Fourier series looks like this:

$$e^{jD\sin(\omega t)} = \sum_{n=-\infty}^{\infty} J_n(D) e^{jn\omega t} \qquad (4.3)$$

The nth Fourier coefficient depends on the modulation index D, and I've called it $J_n(D)$. You may wonder why I've chosen the unusual letter J for the Fourier coefficients; you'll see why in a moment.

The next logical step would be to evaluate the Fourier coefficients by taking the projections of the left-hand side on the basis phasors (as we did in Section 2 of Chapter 7). When you do that you get definite integrals that are frustrating: no matter how hard you try, you won't be able to carry out the integration using any of the tricks you learned in first-year calculus. I say this with some confidence, because the answers are special functions in their own right (always denoted by J_n) that can't in general be expressed using a finite number of any of the functions we've used up to this point, like sines, cosines, or exponentials. The function J_n is called the *Bessel function of the first kind of order n*, and is named after the German astronomer Friedrich Wilhem Bessel (1784–1846).

Fortunately, Bessel functions arise in many areas of mathematical physics, and a lot is known about them. Most mathematical libraries have subroutines for calculating them, and it was therefore easy for me to generate plots of a few of them, which are shown in Fig. 4.1. The next thing we're going to do is use this picture to explain how the spectrum of a carrier is affected when it is frequency-modulated by a sine wave. This will help us understand why FM has been so widely used in music synthesis. After that, I'll return briefly to the question of where Bessel functions come from, and why they pop up so often in many areas of musical acoustics.

Think of Fig. 4.1 as follows. The D-axis represents the amplitude of the modulating signal x. The curve J_0 then shows, by Eq. 4.3, the amount of the zero-frequency phasor in $e^{jD\sin\omega t}$, and since the carrier is multiplied by this, it represents the amount of the carrier itself at its frequency ω_0.

Imagine now that D, the size of the modulating sine wave, slowly grows in amplitude, starting from zero.[†] The size of the carrier is determined by $J_0(D)$, and it will therefore fade, become zero at some point, grow to a peak considerably smaller than its initial value, fade again, and then alternately become zero and reach peaks of diminishing sizes, as D gets larger and larger.

Consider next the first harmonic. The Bessel function J_1 starts at zero, and so initially there is no first harmonic in $e^{jD\sin\omega t}$. This means that initially there is no component in the modulated signal at the frequencies $\omega_0 \pm \omega$. As the amplitude of x grows, however, a similar pattern of growing and shrinking occurs in those components, except the peaks and troughs are not aligned with the peaks and troughs of the carrier we just observed.

[†] If we increase D too fast, it is no longer legitimate to think of x as a sine wave with amplitude D. "Slowly" in this case therefore means slowly compared to the modulating frequency ω.

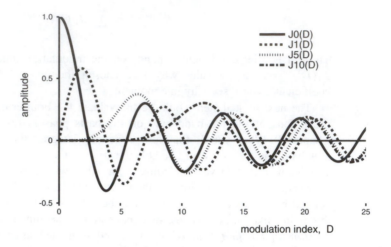

Fig. 4.1 Bessel functions of order 0, 1, 5, and 10. The higher the order, the later the first peak.

The same thing happens with the higher harmonics. For example, J_{10} has its first peak at around $D = 12$, and when the amplitude of x reaches that point the amplitude of the 10th harmonic will peak, producing spectral components in the modulated signal at frequencies $\omega_0 \pm 10\omega$.

The first conclusion we can draw is that the spectrum of the frequency-modulated signal is very complicated, even when the frequency modulation is a single-frequency sinusoid. But we can draw a second general conclusion based on the fact that the higher-order Bessel functions peak at increasingly larger values of D, and this is the hallmark characteristic of FM:

The spectrum of the modulated signal spreads out as the amplitude of the modulating signal increases.

Frequency modulation by a single sinusoid at frequency ω entails multiplying the carrier phasor at frequency ω_0 by a Fourier series with components at integer multiples of ω. Therefore, the only frequencies present in such an FM signal are of the form $\omega_0 \pm k\omega$, for all integers k. In FM radio, the modulating audio signal contains frequencies certainly no higher than 20 kHz, and the carrier frequency ω_0 is typically around 100 MHz, 5000 times higher. The frequency content of the carrier wave is spread out by the process, but it's still accurate to think of the resulting spectrum as being contained in a narrow band around the carrier. (How narrow? See Problem 8.) In sound-synthesis applications, there are no particular constraints on the choice of ω and ω_0. For example, we are perfectly free to choose $\omega = \omega_0$, in which case the FM spectrum consists of components at integer multiples of the carrier, and the FM signal can be thought of as periodic with fundamental frequency ω_0. I ask you to explore other possibilities in Problem 9. Real musical applications in synthesizers or computer programs often involve complicated interconnections of FM signals modulating other FM signals, and designing artificial instruments this way is a highly developed art.

Chowning (see the Notes) gives some recipes for synthesizing a few families of instrumental timbres. To get a brasslike sound, for example, he recommends using equal modulating and carrier frequencies, so that all integer harmonics of the fundamental frequency are present, and then increasing the modulation index linearly during the attack. Figure 4.2 shows the spectrogram of a sound I produced this way. I made the attack longer than recommended so it would be easier to see (and hear) the spectral changes. The spectrum opens up dramatically and the harmonics have momentary dead spots as the modulation index passes through values that make the corresponding coefficients of the Bessel functions zero.

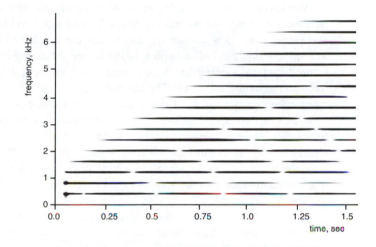

Fig. 4.2 The spectrogram of a synthesized FM sound with fundamental frequency 400 Hz. The modulation index D increases linearly from 1 to 15 over the time period shown. Nulls appear at staggered frequencies in the harmonics. Do you see a pattern?

I'll now return for a moment to Bessel functions. Recall that we began this book by asking where sinusoids come from. We found the answer in the many simple physical systems, like stretched strings and air in pipes, that vibrate in sinusoidal patterns. Bessel functions come up in exactly the same way, except that the vibrating systems are different. The classic example is a planar, circular, flexible membrane constrained only at its perimeter, and otherwise free to vibrate — like a drum head. The wave equation governs the motion of this system, but the geometry leads to Bessel functions instead of sine waves. In fact, it's generally true that any circularly symmetric system obeying the wave equation will lead to Bessel functions. In some sense they are, after real and complex exponentials, the next most natural waveshapes in the universe.

5   The phase vocoder

We now finish with something of a *tour de force*, a practical development from telephone researchers that has also become an interesting tool for computer musicians. It uses many of the ideas we've been studying, including filtering and the FFT, and illustrates the strong connection between the bandwidth of a signal and the amount of information it carries.

The motivation for developing the phase vocoder and similar techniques was largely economic. Telephone companies push many voices through wires, and the more efficiently they can do it, the better the service, the lower the price, and the higher the profits (at least in principle). Today that economic pressure translates to the need for reducing the bit rate — the number of bits per second — required to encode speech so that it can be reconstructed at the receiving end with good quality. Thus, the telephone industry sustains intense research into the nature of speech signals and their efficient representation.

We know we must sample signals like speech at a rate at least twice the highest frequency present, and we must quantize finely enough to keep the quantization noise acceptably low. How can we hope to reduce the number of bits we get when we follow these guidelines? The hope is based on the following intuitive reasoning: speech is produced by moving the lips, tongue, and jaw, and forcing air from the lungs through the mouth and nose. The general shape of the mouth determines, roughly, the overall frequency content of the sound, and that shape is not changing at anywhere near the rate corresponding to the maximum speech bandwidth. To go from one vowel to another, for example, can't take less than, roughly, 1/50th sec, and therefore we should be able to capture the spectral shape with only 100 samples/sec. Put another way, the bandwidth of the signal consisting of the overall shape of the vocal tract has a lower bandwidth than the speech signal itself. We'll call this overall spectral shape the *spectral envelope*.

Besides the spectral envelope of a speech signal, there is also a rapidly varying component: the pitched signal coming from the vibrating larynx for sounds like vowels and broadband noise for consonants. This signal is called the *excitation*. Again, for physical reasons, the excitation can't change its characteristics too fast. The picture developed is that, to a first approximation, speech is produced when a slowly varying excitation signal has its spectrum shaped by a filter representing a slowly varying spectral envelope.

The picture we just developed for a speech signal suggests that if we can extract the spectral envelope and the excitation, we ought to be able to do a good job in transmitting those two signals with fewer total bits than are in the original raw signal. We should then be able to put the two pieces back together at the receiving end to get a convincing restoration of the original speech. This general plan has dominated research in speech representation and compression for at least the past half-century. The phase vocoder developed by Flanagan and Golden is one of the best ways to carry out this program. Let's see how it works.

By now you should (I hope) think of signals as being composed of sums of sinusoids. So it should be natural to break apart a speech signal into narrow-band pieces by using a bank of bandpass filters (see Fig. 5.1). This process should remind

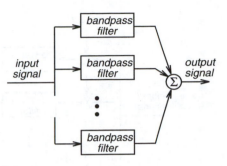

Fig. 5.1 A filter bank. Each channel contains a narrow-band slice of the in-put signal. These pieces can be reassembled to restore the original.

you of Fourier analysis, and I'll return to that idea soon. A phase vocoder uses such a filter bank, where, typically, each filter has a bandwidth of about 100 Hz. A telephone signal, with a total bandwidth of only about 3 kHz, would therefore be broken down into 30 channels, each with a bandwidth of 100 Hz. If we add those signals up, we get back to the original.

Now the narrower the bandwidth of a signal, the closer it is to a pure phasor. The difference is that the magnitude and phase of a narrowband signal change slowly, whereas the magnitude and phase of a phasor are perfectly constant. Intuitively, a narrowband signal can be thought of as the result of amplitude-modulating a pure phasor, as illustrated in Fig. 3.1. The narrower the bandwidth of the original signal, the narrower the bandwidth of the resulting modulated signal. A narrowband signal centered at the frequency ω_0 can therefore be written in the form

$$f(t) = M(t)\,e^{j\omega_0 t + j\phi(t)} \tag{5.1}$$

where $M(t)$ and $\phi(t)$ are narrowband signals representing the (slowly) time-varying amplitude and frequency of the phasor. It is precisely these two signals, $M(t)$ and $\phi(t)$, that we want to capture the spectral envelope and the excitation of a speech signal within the passband of each channel filter.

The next question is how to extract $M(t)$ and $\phi(t)$ from the outputs of the bandpass filters in the filter bank. A good way to see how this is done is to go back to the implementation of bandpass filtering by heterodyning, shown in Fig. 3.3. Figure 5.2 shows the filter bank in Fig. 5.1 with the bandpass filters implemented this way. We'll concentrate attention on a typical channel with center frequency ω_0, but of course you should not forget that there is one branch for each of many channels. This picture has an interpretation we discussed earlier: the multiplication by the phasor $e^{-j\omega_0 t}$ moves the frequency region of interest down to the baseband, the lowpass filter removes everything else, and the multiplication by $e^{j\omega_0 t}$ restores that piece of the frequency content to its correct position along the frequency axis.

The terminology of AM radio is now especially appropriate. The first multiplication by a complex phasor can be thought of as removing the carrier of an AM signal at

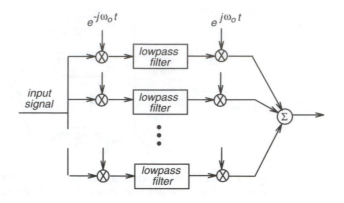

Fig. 5.2 This shows the filter bank with each bandpass filter implemented by heterodyning, as in Fig. 3.3.

the frequency ω_0, and the second can be thought of as restoring it. In between, we detect the modulation, the information-bearing part of the signal. That signal appears as a complex-valued waveform at the output of the lowpass filter.[†] We have in effect built an AM radio receiver, or rather a bank of them, and tuned each channel to a narrowband slice of the original input signal.

All that is left to extract the spectral envelope $M(t)$ and phase $\phi(t)$ is to compute the magnitude and phase of the complex-valued signal appearing at the outputs of the lowpass filters, as shown in Fig. 5.3. If we are interested in bandwidth compression, we can down-sample and transmit the resulting slowly varying envelope and phase signals at a slow rate. Just how low a rate we can get away with depends on how well the signal can in fact be modeled by our spectral envelope/phase picture. We now have a complete phase vocoder.

I should mention that the phase is a very awkward signal to transmit, because it tends to wander off above or below 2π, and in so doing appears to jump suddenly to stay within the usual range between $\pm 2\pi$. It doesn't really jump, of course, but it's difficult to keep track of just what multiple of 2π is included in its true value. In fact, without the jumps it's unbounded, and with the jumps it's not narrowband. (Why not?) The practical solution is to transmit the derivative of the phase, instead of the phase, and then to restore the phase by integrating at the receiving end.

Another important practical point is that the array of signals at the outputs of the lowpass filters can be computed *en masse* using the FFT. If you want to know more about the actual performance of the phase vocoder for data compression, see the original description by Flanagan and Golden referenced in the Notes.

Now I want to describe how the phase vocoder can be used to manipulate sounds in exceedingly interesting ways.

[†] We use complex phasors for the heterodyning, but in practice, equivalent real arithmetic with sines and cosines is used, as suggested by Problem 10.

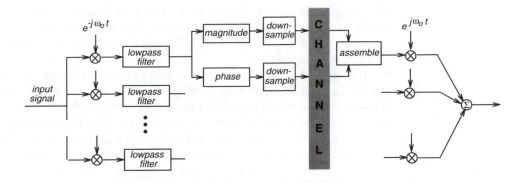

Fig. 5.3 The phase vocoder extracts estimates of the spectral envelope and excitation for each narrowband channel. These are then transmitted at a reduced bit rate. (Actually, the derivative of the phase is transmitted.) The bit rate is reduced by lowpass filtering and decimating, a process we label *down-sampling*.

6 An audio microscope/macroscope

As mentioned, the main motivation for developing the phase vocoder was economic. But science has a way of producing unexpected dividends, and the ability to disassemble and reassemble speech signals makes new things possible. Composers of computer music, of course, were quick to explore the phase vocoder as a way to manipulate all kinds of sounds in artistically useful ways.

Here's a difficult, long-standing problem that the phase vocoder solves, to a practical degree at least. Suppose you have a recording of someone reading a book and it's just taking too long. You'd like to speed it up by a factor of two, say, but you know very well that if you simply play the samples twice as fast, the person reading will be transformed into a chipmunk.

The reason for the chipmunk effect is not hard to find: when speech is played back at faster than normal speed, not only do the words go by faster, but the *frequencies* of all the sounds are increased. At twice normal playback speed, the frequencies are all doubled, which has a drastic effect on how we perceive the sound itself.

The phase vocoder offers an elegant way around this problem, because it decouples the frequencies and the events that make up speech. The idea is to put the pieces of the speech back together so that the frequencies are all half of what they should be. This can be done by simply using half of the phase and half of the original heterodyning frequency ω_0 in each channel. That results in speech in which all the frequencies are half of what they should be, but in which the words go by at the proper rate. If we now play back this new signal at double speed, the proper frequencies will be restored, but now — and this is the point — the words will go by at double speed.

Nothing's perfect, of course, and this method works only as well as the original phase vocoder is successful in separating the slowly varying spectral envelope (which determines the events) from the excitation (which determines the actual frequencies of the sounds). Getting the idea to work in practice usually involves hand-tuning the details of the implementation. In particular, you need to be very careful in choosing the number of channels, and, if the FFT is used to generate the baseband signals, the FFT length and window.

The same idea can be used to slow things down. Just restore the signal with frequencies and phase higher than the original, and then play it back slower. This makes it possible to stretch the sounds in a piano note, say, by a factor of ten, or even a hundred, and new sonic worlds open up. A lot goes on when you play a note on the piano, and this slows things for our aural inspection — a sound microscope. The piece "Still Life with Piano," by Frances White, uses this technique (see the Notes).

It's a good feeling to end with such a happy blend of science and art. In my best of worlds they are hardly different.

Notes

The following gives a wonderfully detailed, first-hand description of the CD digital audio system:

> "Compact Disc Digital Audio," *Philips Technical Review*, vol. 40, no. 6, 1982.

This journal number consists of four articles by Philips staff members, the last of which is devoted to the digital-to-analog conversion process:

> D. Goedhart, R. J. van de Plassche, and E. F. Stikvoort, "Digital-to-Analog Conversion in Playing a Compact Disc," *loc. cit.*, pp. 174–179.

The reverb described in Section 2 is based on the article

> J. A. Moorer, "About this Reverberation Business," *Computer Music Journal*, vol. 3, no.. 2, pp. 13–28, 1979.

Moorer puts the problem in some historical perspective, singling out in particular the pioneering work of Manfried Schroeder in the 1960s. He also gives some good physical motivation for the evolution of his recommended structure, shown in Fig. 2.3.

John Chowning had the wonderful idea of producing sounds using frequency modulation at audio frequencies. His classic paper gives some concrete suggestions for getting started:

> J. M. Chowning, "The Synthesis of Complex Audio Spectra by Means of Frequency Modulation," *Journal of the Audio Engineering Society*, vol. 21, no. 7, pp. 526–534, 1973.

The phase vocoder was first described in the following concise and lucid paper:

J. L. Flanagan and R. M. Golden, "Phase Vocoder," *Bell System Technical Journal*, vol. 45, no. 9, pp. 1493–1509, Nov. 1966.

James Moorer was the first to apply the phase vocoder to musical sounds. The following paper was actually presented at the Audio Engineering Society's conference in
1976:

J. A. Moorer, "The Use of the Phase Vocoder in Computer Music Applications," *J. Audio Engineer. Soc.*, vol. 26, no. 1/2, pp. 42–45, Jan./Feb.
1978.

Details about the practical implementation of a phase vocoder, including a program,
are given in F. R. Moore's book, referenced in the Notes to Chapter 1.

For the example of the musical use of the phase vocoder, hear F. White's "Still
Life with Piano," Compact Disc CRC 2076, Centaur Records, 1990.

Signal processing systems with more than one sampling rate are called *multirate*
systems. The subject is of great practical importance and has a large literature. The
following is a comprehensive book on the subject by prime contributors to the field:

R. E. Crochiere and L. R. Rabiner, *Multirate Digital Signal Processing*,
Prentice-Hall, Englewood Cliffs, N.J., 1983.

The two central problems in multirate processing are raising and lowering the sampling rate of a signal. We've seen examples of both in connection with oversampling.
Decreasing the sampling rate, as we do after oversampled a-to-d conversion, is usually
called *decimation*. Increasing the sampling rate, as we do before oversampled d-to-a
conversion, is usually called *interpolation*.

Problems

1. The 96-coefficient quarter-band digital filter used before oversampled d-to-a
conversion in the CD player is applied to a signal in which three out of every four
samples are zero. Work out the programming details for implementing this filter so
that it makes efficient use of both time and storage.

2. Use a program of your choice to design a quarter-band filter to meet the
specifications shown in Fig. 1.3. When you're done, round off the coefficients to 12
bits and plot the frequency response of the filter with the 12-bit coefficients. How
badly does the frequency response deteriorate?

3. Sketch what Fig. 1.1(e) would look like if it were added to illustrate the effect of
the zero-order hold.

4. The first substantial image of the baseband signal in a CD player after oversampled
d-to-a conversion is well beyond the range of human hearing. Why is it still important
to filter it out? (Hint: It helps to know something about electronics here.)

5. Why is the specified rejection for the quarter-band CD player filter chosen to slope
up as shown in Fig. 1.3?

6. The form of the lowpass comb used in Moorer's reverb filter, shown in Fig. 2.4, is exactly like the plucked-string filter, except the lowpass filter in the loop is feedback instead of feedforward. Is it an accident that these two sound-processing filters have identical structure? Explain if it isn't. Speculate about why a feedback filter is used in the reverb but a feedforward filter is used in the plucked string.

7. Prove that the transfer function in Eq. 2.7, the allpass from Chapter 6 with unit delays replaced by delays of m samples, is also allpass, for all values of m.

8. Commercial AM radio stations in the medium-wave band (around 1 MHz) are usually spaced 10 kHz apart, whereas commercial FM stations (around 100 MHz) are usually spaced 200 kHz apart. Explain how such minimum spacing is determined and why FM stations are so much farther apart.

9. Suppose, as in Section 4, that we use a sinusoid of frequency ω to frequency-modulate a carrier of frequency ω_0. Let the ratio ω_0/ω be N_1/N_2, where N_1 and N_2 are relatively prime integers; that is, all common factors have been divided out. Determine what frequencies are present in the FM spectrum for the cases $N_2 = 1$, 2, and 3. What happens when the ratio ω_0/ω is irrational?

10. Figure 3.3 shows how to use heterodyning with complex phasors to implement a bandpass filter with variable center frequency. Show that the system in the figure below, which uses only real arithmetic, produces the same result, except for a factor of $\frac{1}{2}$. The input signal is assumed to be real-valued.

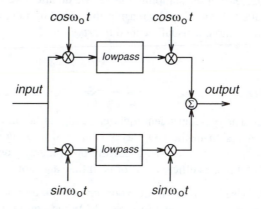

11. I gave the basic reason we might expect to be able to compress a speech signal. Give a similar argument for video signals.

12. Estimate the number of bits stored on an audio CD, using the data in Section 1.

Index